Progress in Bioethics

Basic Bioethics
Glenn McGee and Arthur Caplan, editors

For a list of the series, see page 285.

Progress in Bioethics

Science, Policy, and Politics

edited by Jonathan D. Moreno and Sam Berger

The MIT Press
Cambridge, Massachusetts
London, England

For information about special quantity discounts, email special_sales@mitpress .mit.edu.

Set in Sabon by SNP Best-set Typesetter Ltd., Hong Kong. Printed and bound in the United States of America.

Library of Congress Cataloging-in-Publication Data

Progress in bioethics : science, policy, and politics / edited by Jonathan D. Moreno and Sam Berger; foreword by Harold Shapiro.
    p. cm.—(Basic bioethics)
Includes bibliographical references and index.
ISBN 978-0-262-13488-0 (hardcover: alk. paper)
1. Medical ethics. I. Moreno, Jonathan D. II. Berger, Sam, 1982–
R724.P736   2010
174.2—dc22
                                                                2009015646

10 9 8 7 6 5 4 3 2 1

Contents

Series Foreword

I am pleased to present the twenty-sixth book in the series Basic Bioethics. The series presents innovative works in bioethics to a broad audience and introduces seminal scholarly manuscripts, state-of-the-art reference works, and textbooks. Such broad areas as the philosophy of medicine, advancing genetics and biotechnology, end-of-life care, health and social policy, and the empirical study of biomedical life are engaged.

Arthur Caplan

Basic Bioethics series editorial board:

Joseph J. Fins
Rosamond Rhodes
Nadia N. Sawicki
Jan Helge Solbakk

Foreword

Audi alteram partem. (Hear the other side.)

This book represents an important contribution to an ongoing intellectual debate in bioethics. In general terms, the debate concerns a portfolio of ethical issues that arise in connection with biomedical research, the treatment of persons in a health-care context, and the deployment of new biomedical technologies. Since these areas continue to develop at an extraordinary pace, there is a great deal to consider. Moreover, since it remains unclear whether we are to look to science, to religion, to philosophy, to literature, or to history for guidance regarding how we ought to live, our consideration of matters new and old on the ethical frontier seem simultaneously more exciting, more ominous, and more complicated than ever. Thus I believe that thoughtful and informed debates on these and associated matters are more crucial than ever. This volume, therefore, "enters the lists" at an opportune time.

In areas characterized by continued ethical controversies, intellectuals and intellectual debates are important to a morally mixed society with diverse conceptions of the good, rival ideas of justice, and no final exclusive authority on what we understand to characterize the good life. Indeed, thoughtful intellectual debates can be one of the primary vehicles through which we gain a growing understanding of our cultural patrimony, as entrenched beliefs are re-examined, better worlds are imagined and, in some cases, new ideas gather support and new policy actions are suggested and evaluated. In short, these debates help us understand aspects of the world we have been born into as well as the world as it ought to be. While it is often charged that intellectuals deal "only" in words, as opposed to actions, their words can and do stimulate changes in our human efforts. In short, intellectuals and the debates they generate are indispensable to the spatial distribution of new ideas which give structure to our evolving lives and communities.

Somehow, we human beings are not prepared, as Harry Frankfurt has observed in his book *Taking Ourselves Seriously and Getting It Right*, to accept ourselves just as we are. We feel the need to direct or redirect ourselves, to evaluate ourselves, and to give our actions some sort of meaning. In other words, to use Frankfurt's phrases, we take ourselves seriously and we are interested in getting it right. As a result, we are constantly engaged in a type of anxious self-oversight. We are always judging and reviewing. This is especially true, or ought to be true, of intellectuals, and it is this sort of anxious oversight that generates important intellectual debates, especially in times of rapid change. While such activities are the source of considerable fretting, they are also the source of some of our most important accomplishments.

Significant scientific advances inevitably bring a great deal of anxiety and change in their wake. In particular, there are the ethical issues involved in deciding how to deploy new human powers and the issues of social justice that arise as a result of a realignment or redistribution of resources, power, and interests that inevitably arise out of the deployment of new knowledge. In short, changes, ethical controversies, dislocations, and an ongoing redistribution of costs and benefits will continue to characterize any society that continues to be characterized by significant transformations of one kind or another. This provides fertile ground for intellectual debates of all kinds. The essays in this volume are put forward in such a context, with a focus on some of the contentious ethical issues involved in contemporary bioethics.

Since the biological frontier is moving at an extraordinary pace, we should not be surprised that there is considerable ethical controversy regarding whether and how we should use our new powers. Not only is such controversy inevitable; it may be a good omen, indicating that society is taking these matters seriously. Indeed, such controversy, providing it is thoughtful and empathetic, is the basic vehicle a pluralistic society has available to help it move toward some resolution of the issues that have arisen. These discussions, if they are serious (i.e., if participants might in principle change their minds), reflect an attempt by all sides to shape a future that is consistent either with their values, or more problematically, with their existing patrimony.

Progressive bioethics is certainly not a new discipline. Indeed, in my view bioethics is not yet a separate discipline at all, since, in its contemporary form, it has had too little time for the self-reflection required for the establishment of a new field of scholarship. Nevertheless, a great deal of valuable scholarship has been produced, and those thinking of themselves as "progressive bioethicists" have a distinctive attitude as they approach new issues. This attitude is a predisposition to embrace change as a vehicle to enhance the social, political, and economic conditions they believe to be appropriate for allowing individuals, and through them society, to realize, in the fullest manner, their individual and joint destinies. While no one favors thoughtless action, or action for its own sake, "progressive bioethicists" would tend to favor, for example, the thoughtful application of trial and error as compared to the precautionary principle. Thus they look to scientific achievements as one of the primary means of improving humanity's condition. Perhaps they are distinguished from others most by their conception of human natures and in particular their belief that humanity is capable of taking greater responsibility into its hands.

One of the most ancient human concerns has been what it means to be human, and what humans should become. Moreover, developments in genetic science have brought renewed focus to what we might mean by natural or unnatural human activity. These are central issues, since how we seek to order our lives, our aspirations, and our society depends, in part, on what we believe about human nature(s).

It would be a mistake, however, to allow progressive bioethics to capture or highjack the rhetorical advantage of designating other views, whether more or less radical, as reactionary or "non-progressive" on the one hand or hopelessly utopian on the other. Such adjectives are entirely too pejorative. However, I think it is fair to say that others, particularly more conservative scholars, tend to have a distinctly different view of our human natures and somewhat less enthusiasm regarding the potential benefits of using new scientific discoveries as a vehicle for taking ever-increasing responsibility for the human condition into our own hands. They tend to worry a good deal more about the dangerously seductive nature of new technology, since there are always hidden dangers in it and by itself technological capability is certainly a slim reed on which to build ethical arguments.

I put aside for the moment those on the more radical and authoritarian left and right who already know what is best for all of us. On both the far left and the far right there are more dangerous intellectual and political forces that desire to somehow cleanse society of the "impurities" of the present or the past or the imagined future. We ought to be conscious of the extremes on either end of the spectrum, where everyone is entirely too sure of what we all need, entirely too confident they know how we all should live, and apparently blind to the inevitable ambiguities that characterize the human condition. In any case, it is difficult to have serious conversations with such groups, if serious conversation involves the possibility that participants might, at least in principle, change their minds as a result of hearing the other side.

It is also important for all sides to recognize that in order to move into the future with confidence we need to deploy both imagination and intellect. If we are to have the full benefit of serious discussions about the future, we must construct places to discuss our future together; then we need to provide space for both our imagination and our intellect to work. Our various conceptions of the good life, or the most important virtues, arise not only from our intellect but also from our imaginations, our histories, our feelings, and our memories. Given the plurality of our backgrounds, our moralities, and our narratives, we need to be acutely aware of the absolute need to hear the other side with thoughtfulness and empathy. Moreover, society must have available a social mechanism by which to adjudicate inevitable differences. In short, we need to acknowledge that there is no prospect of maintaining our liberal democracy if we always vanquish those within our society holding different views and insist that there is only one "good" way to live. This strategy leads to cleansing and its associated excesses. In most cases it is not helpful to demonize one's opponents, or to believe that they are not acting from honorable motives.

"Progressive bioethicists" believe that they are searching for ways in which our expanding knowledge base can be used to improve the human condition. Of course, others with quite different views would articulate, in a perfectly sincere manner, the very same objective. Differences or controversies, therefore, reflect differing perspectives on our basic human natures and a different conception of the benefits of using new scientific achievements to take ever-increasing responsibility into human hands. Indeed, I believe that most controversies in bioethics are really a special form of cultural criticism that divides those with deep reservations about developments in our society writ large and those who have a more optimistic interpretation of both our recent experience and the new possibilities being created on the scientific and technological frontiers.

As others have noted, in bioethics there has been, at least in caricature, somewhat of a divide between those who welcome both change and the increasing human powers and responsibilities it implies and those who focus more on the advantages of preserving important aspects of our world and culture as we have found it. In truth, however, the boundaries between these approaches are quite fuzzy and permeable. That is, individuals often find themselves on different sides of this particular divide, depending on the issue. Moreover, the clash between the "ancient" and the "modern" has a very ancient lineage and probably is an inevitable background to significant change of any kind. In any case, the contributors to this volume provide very valuable additions to the ongoing debates, and their essays should cause all those with deep interest in the topics covered to think once again about whether or not their own views should be modified in some fashion.

Harold T. Shapiro

Acknowledgements

We are grateful to the Center for American Progress (CAP), and in particular to John D. Podesta, who has long understood the role of bioethics in a new agenda for progressive policy in America. CAP has given us and our colleagues a venue for the development of many of the ideas discussed in this book.

Thanks are due as well to the interns who have worked at CAP since the founding of the Progressive Bioethics Initiative there, bright and energetic young people in whose hands lies the future of this work. Special thanks go to Sirine Shebaya and Justin Masterman for their assistance with the preparation of the manuscript.

We greatly appreciate the support of the MIT Press and Clay Morgan, who have given us the opportunity to present these ideas to a wider audience.

Our colleagues and friends at the University of Pennsylvania Center for Bioethics and the Yale Law School have provided a rich and challenging environment for our own continuing intellectual development. And most importantly we thank our families, who have put up with many a discussion of bioethics at the dinner table, offered advice and encouragement, and helped in countless ways throughout the whole process.

Sam Berger
Jonathan D. Moreno

Introduction

Jonathan D. Moreno and Sam Berger

This anthology is intended to address a seemingly simple question: What is progressive bioethics? The answer to that question, however, is anything but straightforward. Progressivism and bioethics are both terms still very much up for debate—the appropriate scope of bioethics and the defining aspects of the progressive movement are hotly contested. This book does not seek to definitively resolve these issues, but rather to offer various perspectives on the topic from bioethicists who consider themselves progressives and those who do not. We aim to begin a dialogue about the nature of progressive bioethics, to provide a foothold for those interested in understanding the field, and to plant our own flag in this fertile but largely unexplored territory.

At its heart, the question of the nature of progressive bioethics lies at the center of two burgeoning phenomena: the entrance of bioethical issues into the realm of conventional political discourse and the rise of self-styled progressives who hearken back to the Progressive Movement at the turn of the last century. To consider this question at all suggests that large changes are afoot, both in the field of bioethics and on the political left— changes that will greatly affect American society as a whole.

It is hard to deny that bioethics has become more political in recent years, shattering what one of us has called the Great Bioethics Compromise (J. Moreno, The end of the Great Bioethics Compromise, *Hastings Center Report* 35, 2005, no. 1: 14–15), which might be summarized in the following formula: Keep a close eye on scientific innovation for its societal implications, apply the brakes now and then as needed through regulations or guidelines or just the glare of public discussion, and let the bioethicists be the ones to analyze how all this is going. The Great

Bioethics Compromise was an implicit attempt to hold in balance two facts otherwise in tension: that there were from the beginning serious disagreements about the implications of certain technologies, especially in the area of reproduction, and that it was important to retain academic politeness within the small fraternity of bioethicists in the early years.

In spite of some bumps in the road during the 1980s and the early 1990s—the Baby Doe controversy, the discussions on human embryo research, a federal bioethics commission that failed to get off the ground as a result of disagreement about members' positions on abortion—the agreement to remain civil and outside the larger political currents held. The cloning of Dolly the sheep in 1996 and Jamie Thomson's landmark paper on human embryonic stem cells stimulated a change in the atmosphere, however, and long-standing issues in the bioethics community finally boiled to the surface. Meanwhile, social issues were taking center stage in the political realm, and the Bush-Gore election debacle aggravated tensions throughout American society, including academia. When President Bush appointed a bioethics council widely perceived to be intellectually if not numerically tilted to social conservatism and headed by a prominent traditionalist, the die was cast.

It is commonly noted that the membership of the President's Council on Bioethics included such unusual appointments as that of a nationally syndicated conservative newspaper columnist. But perhaps it was more significant that, during the first Bush term, the council's senior staff was to a great extent drawn from a conservative Washington think tank, and that these same individuals were editors and writers for a journal published by that institution. These writers were young, smart, and had a good understanding of the political process and the making of public policy. Importantly, they also had not come through the customary pathway of graduate bioethics training with at least some exposure to the clinical setting. They were largely political philosophers with a special interest in bioethical issues. They came to Washington with more than just an ideology, however; they came with an agenda.

This new generation of conservative bioethicists began to advance a relatively coherent view of bioethics, one that focused on the dangers of science and the downfalls of our efforts to improve the world. Perhaps most important, these bioethicists provided a road map for social conservatives less familiar with bioethics, helping them to articulate bioethical policy positions in persuasive ways and to understand new bioethical issues in the context of long-standing social debates. The coherence and message discipline that was a hallmark of the right had finally made its way to bioethics, with impressive political results at the local, state, and federal levels.

Bioethics suffered from the same problem as the rest of the political left: although academia was dominated by leftist views, it had not succeeded in producing a younger generation of savvy, policy-oriented thinkers. Nor had it done an effective job of connecting bioethicists with the larger movement and message apparatus of think tanks, journals, and elected officials. Despite its prevalence within the university walls, left-leaning bioethics was losing in the political realm. The public policy conversation was clearly being driven by conservative thought.

In an effort to correct the ideological imbalance in the public sphere, the last few years have seen the rise of a new progressive movement, one that draws ideas and people from both leftists and moderates. Rather than continue the tired debates of the past, progressives seek to deal with the new and difficult problems confronting our country. Driven by institutions like the American Constitution Society and the Center for American Progress (a think tank where we both worked), the progressive movement has been increasingly politically successful, culminating in the election of Barack Obama. Obama is a truly progressive politician, as willing to throw aside the failed orthodoxies of liberalism as those of conservatism, and with a deep respect for empirical facts and expertise. The next few years thus represent a unique opportunity for progressives, including in bioethics, to articulate their beliefs to a welcoming audience and make the case for progressivism to the American people. We hope that this book will serve in some small way in this task, helping bioethicists and the lay public alike to better understand what progressive bioethics is, and why in the current political climate it is so sorely needed.

The chapters in part I seek to explore the meaning of progressive bioethics itself. In chapter 1, we discuss progressivism and the close ties between this political theory and bioethics, then suggest characteristics of a progressive orientation toward bioethics. In chapter 2, Richard Lempert argues that progressive bioethics is needed, but it must seek to be non-ideological, wedded to basic principles of sound ethical inquiry that eschew agenda-drive conclusions.

Part II turns to the current politicized state of the field. R. Alta Charo examines the rise of conservative bioethics and its philosophical foundations, then suggests that progressives should instead ground their beliefs in Enlightenment philosophy and support for free scientific inquiry. Kathryn Hinsch looks at the institutions and methods that have proved useful to conservative bioethics and discusses how progressives can learn from their success. Laurie Zoloth argues that conservative bioethics focuses on the wrong questions, and outlines what she considers a progressive set of issues for bioethics to address.

Part III provides insight into the interplay of progressive bioethics and major aspects of the field: professionalization, national commissions, and religion. Paul Root Wolpe writes about the move toward professionalization in bioethics, and what this should and could mean for progressives. John Evans examines the sometimes problematic relationship between bioethics and religion, suggesting ways progressives can reduce the tension and counseling against ignoring the question of ends in the future. Eric Meslin considers whether progressives should create a new model for bioethics commissions that is better suited to creating real change.

Part IV addresses an area of particular contention among progressive bioethicists: the appropriate stance toward biotechnology. James Hughes argues that progressives committed to individual freedom and equality should be largely in favor of new biotechnologies. Marcy Darnovsky, drawing on the lessons from the environmental movement, suggests a somewhat more cautious approach.

Part V explores whether the negative effects of politicization can be mitigated by finding common ground. Arthur Caplan describes the politicization of bioethics; he argues that there is no way back, and that this is not necessarily a bad thing. Michael Rugnetta explores how a progressive Catholic can approach the issue of conscience clauses. Daniel Callahan suggests that to answer a big question like how to fix America's failing health care system we must turn our attention away from competing ideologies and toward determining the shared end we wish to achieve. Bill May, a former member of the President's Council on Bioethics, reflects on the opportunities to find common ground between bioethicists of all persuasions and ideologies.

Progress in Bioethics

I

Bioethics as Politics

1

Bioethics Progressing

Sam Berger and Jonathan D. Moreno

The last few years have seen increases in the numbers of self-described progressives and progressive organizations. The new progressive movement is not simply an attempt to abandon the politically compromised label 'liberal', but is rooted in certain commonalities between the turn of this century and that of a hundred years ago. Progressives in the early twentieth century faced economic and political changes that seemed to threaten their values and even their very way of life. Similarly today, globalization weakens the ability of the nation-state to regulate its economy, and the global security landscape is dotted with more and more powerful and unpredictable actors, even as we become increasingly aware of growing wealth disparities and of our interconnectedness to people on the other side of the world. The effects of these changes have been felt acutely in the United States; people are concerned about losing jobs in an international marketplace, about the failing social safety net, and about security threats from abroad. The realization that new solutions are needed to confront these issues and many others underlies the modern progressive movement.

In this chapter we will place contemporary American progressivism in historical context and identify some central themes of progressivism in bioethics. As a political philosophy, progressivism complements and builds upon the values of liberalism by emphasizing transparency as an important element of democratic processes, the importance of expertise rather than ideology in public policy, public regulation of concentrations of wealth and power, and activism in the pursuit of social justice. American progressivism is also deeply influenced by pragmatism, which encourages a focus on achieving the goals of social policy by means that work rather than on ideologically based *a priori* judgments about the efficacy of governmental or private sector solutions to public problems. A progressive approach to bioethics is particularly appropriate because, as we shall explain, the core idea of progress is closely associated with the modern idea of science, and bioethics itself embodies a progressive spirit.

As progressives who work in the field of bioethics, we are interested in the ways these values may guide policies and practices in the life sciences, as well as in pitfalls progressive bioethicists must be careful to avoid. Just as America faces major economic and political changes, the concerns stimulated by modern biology are in important respects unprecedented, and are having a significant effect on Western society as a whole. The cloning of Dolly the sheep, the isolation of human embryonic stem cells, and the mapping of the human genome have catapulted biology, and thus bioethics, onto the national stage. One of President George W. Bush's first major policy initiatives concerned federal funding for stem cell research, and that issue has been prominent in a number of recent elections. Bioethics has also become a major concern of the conservative movement, particularly among neoconservatives and the religious right (Hinsch 2005). But the need for a progressive bioethics extends beyond simply a political response to the burgeoning public interest in the field. The progressive focus on bioethics represents something greater: a realization that these are significant changes in the life sciences, changes that progressives would do well to pay attention to and address.

While conservatives have devoted significant attention to biotechnology issues, their approach has not been successful. Although admirable in urging a focus on broad philosophical questions, in practice the conservative response to legitimate issues raised by biotechnology has often been ineffectual and shortsighted. Conservatives frequently default to reflexive opposition to new technologies, an opposition that is almost always overtaken by practical events. Worse yet, this all-or-nothing approach leaves them, and the rest of society, unprepared to address the challenges that are raised by the gradual adoption of new technologies. Consider conservative opposition to *in vitro* fertilization, which culminated in President Ronald Reagan's decision not to fund IVF research. Rather than smothering the nascent technology, his actions simply allowed it to develop in the private sector, unregulated and self-regulated, which has caused a number of troubling developments we are now beginning to face (Mundy 2007). By preventing the government from regulating cloning and stem cell research through federal funding and other vehicles of public control, research opponents actually help foster similar problems.

Rather than seek to stop change in its tracks, progressives have historically understood that one can view aspects of it as problematic without rejecting it wholesale. Seeing rapid industrialization in the beginnings of the twentieth century, progressives recognized that it was leading to a dangerous accumulation of political and economic power in the hands of a few. But they also saw that it was strengthening the American economy, and that it could not be stopped or reversed. Rather than fight economic and political change in principle or accept it as inevitable, progressives sought to ensure that this change would be influenced and constrained by widely shared values of rewarding hard work, providing economic opportunity, and strengthening democracy.

The changes wrought by biotechnology are in a similar vein—challenging, but potentially very beneficial—and thus require a similar approach. Progressives should not unduly oppose the use of new technologies, but they also should not forget that these technologies must be controlled and regulated so as to comport with our shared values. There is no denying that many find the implications of new biotechnologies disconcerting, and for good reason. Despite their tremendous promise to improve our lives, they also present novel and sometimes unsettling prospects. The synthetic manufacture of deadly pathogens, growing disparities in access to both conventional and newly developed medical care, and the continued commodification of the body are all issues posed by recent medical discoveries. Progressives must be careful to balance enthusiasm for the positive possibilities of new biotechnologies with a healthy respect for their potentially negative effects.

Further, many of these issues cannot be easily addressed with old methodologies. Consider, for example, the tensions facing the reproductive rights movement's support of reproductive choice when new biotechnologies offer (or will offer) the potential not only to choose when to have a child, but also choose what characteristics it will have (Berger

2007). The issue of reproductive choice takes on new meaning when people can use it in a discriminatory manner, choosing not to bear a child who is a girl or who has a disability. What was once a rallying cry for reproductive rights now seems problematic in certain contexts. These types of novel problems will require equally novel and adaptive solutions. As progressives, we must understand that change will come; the question is what we will do to shape that change.

Progressivism: Past and Future

We can learn much from the original progressives of more than 100 years ago. The Progressive Era was a time of optimism as well as rapid change. Many took seriously the proposition that government could be an institution of reform. At the dawn of the twentieth century, America was feeling the effects of the Gilded Age, a second industrial revolution that made local businesses give way to factories and moved political power from the small towns to the big cities. A small group of businessmen capitalized on these changes, amassing vast wealth and power. With these changes came new problems as ordinary people struggled to adjust to the changing economic realities and government was corrupted by the powerful.

Seeing these problems, and believing in their own ability to fix them, Progressives embarked on one of the most ambitious plans for reform since the Founding. They sought to embrace the potential of these new developments, but also to ensure that this potential was shaped by their shared values. "Progressives arose to bring order to both politics and the economy, but their quest was at least as much about morality as about political economy. The values of the Progressives were rooted in the old virtues, even as they accepted that the tides of industrialism could not be turned back." (Dionne 1996b, 35) Muckrakers like Upton Sinclair exposed the deplorable working conditions in factories, activists like Jane Addams provided help to the poor immigrants who flocked to urban areas looking for work, and trustbusters like Theodore Roosevelt broke up illegal business monopolies and heightened industry regulation. Academics also joined the fray, realizing that "the function of 'social science' wasn't simply to dissect society for non-judgmental analysis and academic promotion, but to help in finding solutions to social problems" (Moyers 2003). Progressives determined that to limit the deleterious effects of national corporations they would have to give increased power to the national government to protect the interests of ordinary people.

In order to effectively use government as an instrument of the popular will and public interest, Progressives first had to clean it up, rooting out corruption and replacing it with competence and a spirit of public service. Newspapermen like Lincoln Steffens exposed government corruption, while Progressive politicians such as Robert LaFollette and George Norris ran for public office and won. They created fairer labor standards, publicly owned or regulated sanitation, transportation, and utility systems, and better consumer protection. They passed laws barring corporate campaign contributions, and a constitutional amendment allowing the direct election of senators in an effort to clean up the "Millionaire's Club" the Senate had become. They worked to make government transparent, accountable, and fact-based.

Yet the Progressive Era was not an unqualified boon, particularly in the realm of bioethics. While everyday workers saw improvements in their lives, there was also a resurgence of racism and segregation, encouraged by a "science [that] increasingly endorsed many Americans' belief that some races were better than others" (McGerr 2003, 192). And the same impetus that inspired academics to fight for workers' rights also caused some of them to associate themselves with eugenics and social Darwinism. Tainted as it is with the horrors of Nazi Germany, it is hard to imagine eugenics as a progressive movement. Yet eugenics was widely viewed as the progressive biology of the day, justifying a public policy that included the forced sterilization of "inferior" people such as the mentally retarded, the deaf, and certain ethnic and racial groups. No less a progressive leader than President Theodore Roosevelt said "We have no business perpetuating citizens of the wrong type." (Moreno 2007) Although progressives were not alone in embracing eugenics, and although some of them were among its toughest critics, progressivism must bear its share of the blame for attempting to elevate a biological theory to pubic policy.

The resurgence of progressivism has focused on emulating the economic and political successes of the era while avoiding the moral failures.

Contemporary thinkers liken the economic dislocation caused by the rise of transnational corporations and international competition to the problems caused by national corporations in the 1900s, which resulted in similarly dramatic changes in political systems and moral values (Dionne 1996b). These new progressives have taken up the challenge of ensuring that the economy still has a place for the individual, that politics does not become a plaything of the powerful, and that we do not lose our sense of common purpose and values in the face of tumultuous change (Halpin and Teixeira 2006).

But we also face an era of unprecedented biological change. Scientists continue to develop the ability to clone mammals, regenerative medicine seeks to unlock the power to heal ourselves with our cells, synthetic biology may allow us to create new species, and genetic modification offers the potential to radically alter our DNA. As scientific changes challenge and revise our very definition of life (Borenstein 2007), there is a special role for a progressive bioethics. In working to ensure that these changes improve the common good, we must look to hard-won values of respect for persons and protection of human dignity.

Science and Progressivism

A distinctly progressive bioethics is a natural outgrowth of the close connection between progressivism and science. That the words 'progressive' and 'science' gained their modern meanings at the same time was no historical accident, but rather a demonstration of the shared belief inherent in both the scientific method and the notion of progress that the power of knowledge acquired through systematic inquiry can improve the conditions for human flourishing. In many ways, progressivism is simply the application of the method of science to the development of public policy. At its core, "the American Progressive tradition [is] resolutely experimental rather than reflexively ideological, in constant search of new methods, insistent on continuous reform" (Dionne 1996b, 15). Progressivism is predicated on the questioning of assumptions, on openness of inquiry, on reliance on empirical data, and on transparent communal investigation. Policies are to be based on experiment, not belief; Justice Louis Brandeis expressed this attitude when he referred to the states as laboratories of democracy in which new policies could be tested and improved. Just as good science requires a community of informed and empowered researchers reconsidering existing assumptions in light of data, good government requires knowledgeable and capable citizens and legislators doing the same.

The application of scientific principles to governance can be seen not only in the style but also in the substance of progressive policymaking. Progressives brought regulators with greater scientific expertise and adherence to the scientific method into government to ensure food and drugs were safe. They sought to bring empirical analysis to the selection of members of government itself, greatly expanding the number of government jobs given through the merit system. And they sought to replace the old system of nepotism and corruption that had dominated politics with a government that was responsive to hard facts instead of cold cash. Progressives saw the scientific method as a singularly successful mode of objective inquiry, one that would have similar successes when applied to government.

Progressivism's Promise and Perils

Progressivism is as promising an approach to governing today as it was a hundred years ago. Embracing the spirit of American pragmatism, progressivism focuses on results rather than ideology. Thus, certain problems may require more government intervention, while others may be more amenable to market solutions; only data and experience can determine which is the best means of addressing an issue. Progressives also acknowledge the changing nature of society, understanding that past solutions may not be applicable to current problems. More than that, progressives are hopeful, believing in the capacity of human beings to shape a better world for themselves and their children. They appreciate the possibilities of technology and the use of scientific and technological expertise to achieve our most ambitious goals. Rather than opposing change, progressives embrace the possibility of a new world, seeking to shape it through our shared values into the world we want it to be. And progressivism is mindful of the less powerful, the people in whom political power is and should be vested, ensuring that government works for their interests. This flexible, evolving philosophy is well suited to a quickly changing political landscape in which yesterday's wisdom is today's folly.

Like any governing philosophy, however, progressivism has the potential to go awry. Exaggeration of progressives' support for expertise, belief in our ability to effect positive change, or concern with practical results can lead to impulses that are antithetical to the spirit of progressivism. None of these problems are necessary results of a progressive sensibility, but they do point to potential pitfalls within progressivism that could subvert its effectiveness and its ultimate goals. To avoid these dangers, we must understand the excesses from which they stem.

The belief in the potential of science, evidence, and expertise to solve societal problems can lead to dangerous elitism if not balanced by a keen sense of the limitations of science, as is true for any human endeavor. When relying on expertise and evidence, there may be a tendency to ignore people who are not experts, or to discredit arguments that rely more on moral sentiment than on science. Of course, democracy is based on the notion that individuals can and should make their own political decisions, and creating an environment that allows an engaged citizenry to make these decisions is at the heart of the progressive project. Yet those who strongly support rational, fact-based reasoning may shy away from the messiness of democratic consensus building and gravitate toward the seemingly clearer and cleaner world of elite decision making. This strain of elitism coupled with a belief in the objective truth and power of science was apparent in some progressive reformers in the 1960s, including overzealous social scientists who overestimated their ability to repair social ills. Their perceived failures to fix problems such as de facto school segregation and poverty led to a backlash among intellectuals who "doubted that imperfect and unpredictable human beings could be organized socially on the basis of 'scientific' knowledge alone" (Dionne 1991, 60). Many of these critics coalesced around the new journal *The Public Interest*, which aimed to show that the capacity of social policy to fix intractable problems was limited; eventually, many of those critics would become major figures in the neoconservative movement. Conservatives trace progressive elitism directly back to the optimism of the Progressive Era. The noted conservative intellectual William Schambra erroneously describes Progressives as those who wanted the "transfer of political power away from everyday citizens and their chaotic, parochial, benighted local organizations, often steeped in foolish religious mythology" and who believed that "power should instead be put into the hands of centralized, professionally credentialed experts trained in the new sciences of social control" (2006, 2). While Schambra vastly overstates the undemocratic impulses in progressivism, his critique shows the concern many have for advocating too strongly for expert leadership, particularly in areas where expertise itself is lacking.

Elitism can also lead to a broad utopianism when progressives become too sure of their ability to address any and all of the world's problems. Not recognizing human limits, they may overreach and fail, thus doing more damage than if they had done nothing. The historian Michael McGerr view today's impoverished politics as a direct result of the excesses of the Progressive Era, which "offer[ed] the promise of utopianism—and generat[ed] the inevitable letdown of unrealistic expectations" (2003, xiv). Try to do too much and you may convince people they are incapable of doing anything. But the danger lies not only in lowered individual expectations, but also in unexpected negative results from large-scale, poorly understood changes. The neoconservatives termed this the "law of unintended consequences," arguing that in the 1960s "one well-intended program after another had failed, often by solving one problem only to create another one" (Dionne 1991, 60). In abandoning tradition, conservatives argue, progressives revealed a hubristic belief that they knew better than centuries of painstakingly accumulated human knowledge. Seeing the problems too narrowly, progressives failed to understand the totality of their actions, and at times did more harm than good.

And there is some reason for present-day progressives to be wary of overstepping their bounds. As evidenced by the Progressive Era, the scientific impulse can be abused. Faith in progress inspired protection of workers and accountable, open government, but also was twisted to support eugenic sterilizations stemming from pseudo-science. As long as science carries cultural cachet, people will attempt to justify terrible actions by recourse to it. And there are always limitations to our knowledge and our ability to effect change; the Green Revolution and the war in Iraq are clear testaments to that.

There is also concern that progressives, in their constant efforts to achieve reforms and to solve one problem after another, will lose a sense of the larger picture. Rather than think about the ends, progressives could find themselves too caught up in the means, focusing on the most efficient way to achieve a goal without adequately questioning whether it is desirable. Anticipating the neoconservative critique of mainstream bioethics, this concern was forcefully articulated by Randolph Bourne, a progressive who split with John Dewey over Dewey's support for American involvement in World War I. Criticizing pragmatists, the philosophical school most closely associated with progressivism, Bourne complained they were elitists who were "hostile to impossibilism, to apathy" and had thus created a generation of young intellectuals "immensely ready for the executive ordering of events, pitifully unprepared for the intellectual interpretation or the idealistic focusing on ends" (1967, 88). The young pragmatists had "absorbed the secret of scientific method as applied to political administration . . . [but] they had never learned not to subordinate idea to technique" (ibid., 88–89). This concern seems surprising in view of other commentators' suggestions that Progressives were prone to utopian idealism, but it demonstrates that excess has the potential to corrupt underlying goals in many different ways.

There are strong echoes of these worries in the conservative (and especially neoconservative) bioethicists' project to return to the bioethics of the late 1960s and the early 1970s, when the conversation focused on the moral ends of the life sciences rather than on the appropriate means of utilizing technologies. They worry that in practice bioethical theory has become "thin" with its emphasis on personal autonomy, rather than "thick" with reflection on the goals of medicine, the nature of humanity, and the preservation of human dignity. In particular, Leon Kass, the former chairman of George W. Bush's bioethics council, has called for a "richer" bioethics that does not have moral consensus as a goal and seeks more than mere procedural solutions to ethical dilemmas. These conservative bioethicists fear that the lack of concern for ends could leave progressivism lifeless, a mechanical pursuit of one goal followed by the next, with no uplifting moral vision. In a sense, Bourne and others worried that Progressives' focus on science was causing them to forget the poetry, the humanity of their pursuit. Of course, these concerns are directly connected to worries about elitism and anti-democratic impulses stemming from a distancing of professionalized reformers from the "irrational" masses. Their present-day form can be seen in conservatives' efforts to paint their opponents as "pointy-headed intellectuals" too caught up in the ivory tower of academia to understand ordinary people, and in the left's perpetual worry that it is so focused on policy prescriptions that it lacks a narrative or vision with which to connect with the common man.

It is no accident that the themes of elitism, utopianism, and excessive concern with means also appear in conservative critiques of science, particularly in the writings of neoconservative bioethicists. Drawing on the distrust of earlier neoconservatives for social science expertise, this later generation has exaggerated their worries to encompass the hard sciences as well. The theme of arrogant scientists, dismissive of the common people and unwilling to accept any of their restraints, is a common one. It stretches at least as far back as Mary Shelley's *Frankenstein*. The neoconservative bioethicist Eric Cohen captures the critique in its latter-day form:

From the beginning, science was driven by both democratic pity and aristocratic guile, by the promise to help humanity and the desire to be free from the constraints of the common man, with his many myths and superstitions and taboos. The modern scientist comes to heal the wretched bodies of those whose meager minds are always a threat to experimental knowledge. (2006, 27)

This rhetoric has been extremely strong in the stem cell debate, in which researchers are frequently described almost as if they are mad scientists operating outside the norms of society (Moreno and Berger 2006). Worries that scientists are playing God are partially concerns that they lack a sense of humility—the understanding of human weaknesses and limitations that is the hallmark of conservative thought. Neoconservative bioethicists believe that "humility, alas, is not always a prominent scientific virtue, at least among the most prominent scientists, and especially among many modern biologists" (Cohen 2006, 27).

And neoconservative critics of science also argue that scientists are so caught up in gaining knowledge about the world that they do not stop to think about the morality of their actions, or understand that science itself cannot answer those moral questions. Far from seeing science as value neutral (a widespread view among the general public), neoconservatives see it as value laden in a way that excludes many moral ends. Yuval Levin comments: "In forcing the world into this [scientific] form, science must necessarily leave out some elements of it that do not aid the work of the scientific method, and among these are many elements we might consider morally relevant." (2006, 35) Leon Kass goes even further, arguing that "modern science rejects, as meaningless or useless, questions that cannot be answered by the application of method" (1993, 8). Echoing Randolph Bourne, Kass describes a cold, remorseless science reminiscent of the opening pages of Aldous Huxley's *Brave New World* (a neoconservative favorite), asserting that "the so-called empirical science of nature is, as actually experienced, the highly contrived encounter with apparatus" and that "nature in its ordinary course is virtually never directly encountered" (1993, 7). These neoconservatives argue that scientists are too far removed from the world to properly consider the effects of their actions.

Neoconservatives see science and progressivism as so intertwined that the two are often combined in their minds, concerns about hubristic scientists mingling with old animosities toward leftist reformers. William Kristol (son of neoconservative founding father Irving Kristol) and Eric Cohen made the connection explicit in their discussion of therapeutic cloning supporters, describing them as "an odd mixture of the hubris of the medical researcher seeking to lead his fellow men beyond nature, and the sentimentality of the post-Communist romantic, who sees in genetic science man's new hope for building a kind, just and liberated heaven on earth" (2002, 300). For neoconservatives, these impulses are one and the same, and both must be vigorously opposed.

Bioethics as a Model of Progressive Public Policy

Bioethics has a close connection to pragmatic progressivism, and is in many ways a progressive strain of the existing medial ethics community. The modern bioethics movement was in part a product of the human rights fervor of the later 1960s, and in part a result of a small group of thinkers' concerns about the implications of the biological revolution that could be glimpsed over the horizon. What began as an academic conversation about genetic modification, reproductive technologies, replacing organs, sustaining life, and conducting human experiments quickly became a matter of law and public policy.

The new bioethics was distinct from traditional medical ethics not only in the problems it confronted, but also in its emphasis on the rights of patients and their families to make crucial decisions that historically had been made by physicians. In this respect it dovetailed with a growing public desire to open up medical decision making and to vest greater authority in the individual. Institutionally, the movement to create ethics committees at hospitals caught fire after the Karen Ann Quinlan case. Ethics committees represented a practical alternative to legal action when cooperation and communication between patients and caregivers broke down, or when the medical issues were of unfamiliar complexity. In acute clinical situations, academic theories have little leverage; the ethics committee process represents pragmatism at the "micro" level of the individual case (Moreno 1995). Thus, bioethics embodies such progressive values as pragmatic problem solving and the desire to make large, impersonal institutions more responsive to individuals.

At the "macro" level, no academic field has been so closely identified in its early and continued development with governmental advisory bodies as bioethics. Beginning with the National Commission for the Protection of Human Subjects of Biomedical and Behavioral Research in the 1970s, and continuing through the President's Council on Bioethics, these panels have attracted a great deal of attention from stakeholders and in some cases have created lasting policy frameworks, including regulations on the use of humans as research subjects, standards for informed consent, criteria for review of human genetic testing, rules on human reproductive cloning, and conditions for research involving human biological materials. To be sure, many of the proposals of these bodies had little or no influence, especially if they seemed to go beyond the readiness of the political system to accept them. From a sociological standpoint, it is important to note that the legitimacy of bioethics as a field was partly conferred by a series of bioethics commissions created by both Democratic and Republican administrations. These panels have not been limited to the presidential level; numerous bioethics advisory committees on special topics, including oversight of recombinant DNA experiments and human testing, have been created by various cabinet-level agencies. This reliance on advisory commissions reflects a progressive sensibility; President Theodore Roosevelt appointed the first presidential advisory committee. More important, the systematic approach taken by commissions—engaging knowledgeable experts in the process of assessing the problem at hand, hearing the views of various stakeholders, gathering evidence, and proposing new policy options—has come to be closely identified with progressive policymaking.

Of course, the pragmatic approach to bioethics has often been criticized as "instrumentalist" or merely means-oriented, and as too easily lapsing into a "thin" discussion about process rather than a "thick" discussion about ultimate goals and moral purposes (Evans 2002; Callahan 1996). Although we grant that in the public policy sphere important ethical considerations can too easily be swallowed up by procedural concerns, we reject the view that serious moral reflection about ends as well as means is incompatible with crafting public policy. The goal of making human beings more fully voluntary participants in research, for example, is embodied in the requirements for informed consent. The value of safety in studies of drugs and devices is realized in formal risk-benefit analysis. And, most important, progressivism respects the foundational value of a liberal polity that individuals should have maximal freedom, consistent with the public interest, to pursue their own vision of the good life. What the critics of bioethics view as succumbing to procedural norms, progressive bioethicists see as efforts to reconcile widely held values with respect for individual rights in a pluralistic democracy.

The justifiable criticism and self-criticism of the bioethics movement should not obscure the constructive innovations associated with the social practice of bioethics. Bioethicists have institutionalized the agenda of patients' rights in the clinical setting, and they continue to raise concerns about patients' rights. It would now be unthinkable—and contrary to professional and regulatory standards—for a hospital or a health-care organization to lack a mechanism for addressing ethical issues. Prior review of human research protocols by an ethics board is now a well-established international norm. An international community of scholars has recently launched an energetic movement to address questions of global ethics. And although neoconservatives have urged a thicker rationale for prohibiting human reproductive cloning beyond the risks to the fetus and the mother, the fact remains that the bioethics and life sciences communities have argued for a ban on attempts to clone a human being.

The social practices of bioethics therefore embody core progressive principles: that progress is possible, that pragmatism should prevail over ideology, that both individual rights and the common good can be respected and promoted, and that sound public policymaking requires a respect for evidence and a willingness to change familiar ways of operating. In the coming years, these principles will be needed to address challenges that arise from advances in personalized and regenerative medicine, stem cell research, synthetic biology, and neurotechnology.

In order to adequately address the new biological changes that will be important issues in the twenty-first century, progressive bioethics must reflect the best in both progressivism and bioethics. It must retain progressivism's optimism and drive, and it must retain a belief in the capacity of individuals and government to work together to solve even the largest of problems. Rather than shy away from challenges or hearken back to a bygone era, progressive bioethicists should engage with the world as it is and contemplate how it might be. But this optimism must be tempered by the lessons of the past and by an understanding of the complexity of many of the social and scientific problems that confront us. Progressive bioethics suggests a cautiously optimistic approach to science that acknowledges uncertainties but is not paralyzed by them.

Progressive bioethics must also remain non-ideological, unwedded to certain policy prescriptions as absolute truths, and not distracted by silver bullets that promise much but deliver little. It should be grounded in the best empirical evidence, staying true to the scientific method. Yet this lack of ideology should be restricted to means. Progressive bioethicists must maintain a clear sense of how they want the world to look, and why they want it to look that way. Furthermore, these goals must be subject to debate as new technologies or other changes alter the world we live in. The answers to these moral questions are not found within science, but they help to define science's appropriate goals. Progressive bioethics must ensure that the excesses of science do not threaten society's core values and must remember that the shared moral concerns of the community are as important as medical, economic, and political concerns.

Progressive bioethics recognizes the importance of expertise, particularly in science, and the capacity of a community of experts to be self-correcting. Just as transparency creates the best public policy and ensures accountability of leaders, so too does the openness of the scientific community allow for the best theories to be revealed through constant testing and challenge. Progressive bioethicists must ensure that this openness and accountability remains ever-present in expert communities. But they should also ensure that the limits of expertise are recognized. In a pluralistic democracy, no individual is an expert on how others should lead their lives or how society should be shaped. These determinations are instead made by the people as a whole through their individual everyday decisions, through their elected leaders, and through opportunities to participate in the public dialogue on issues of import. It is crucial that progressive bioethics ensure that opportunities for dialogue continue, whether through the presence of citizens on ethics panels or through active engagement with the citizenry on bioethics issues (as is currently being attempted in Britain in a variety of ways).

Progressive Values in Bioethics

Describing progressive bioethics through a series of specific policy decisions is next to impossible. Difficult with any overarching theory, it is even more so with one that is explicitly non-ideological and wedded to the fruits of inquiry with respect to particular cases. But it is possible to describe the sensibility of progressive bioethics—that is, the types of concerns that can and should inform bioethical discussions. The four major values of progressive bioethics are *critical optimism, human dignity, moral transparency,* and *ethical practicality.* These values do not lend themselves to any specific policy in perpetuity, but serve to shape the debate in ways that are evident in today's discussions.

Critical Optimism

Progressive bioethicists are critical optimists. They understand the tremendous potential of science and technology to improve our lives and our world. They recognize the benefits of penicillin, automobiles, and assisted reproductive technologies. But they also understand that technology and science are not unqualified goods. They remember the lessons of the atomic bomb and of dangerous human experimentation. Thus, they are oriented toward the potential good of the future, believing that on the whole science and technology have been extraordinarily positive forces in our world. Yet they do not forget that this is due in part to our efforts to constrain and shape those technologies, and, at times, to prohibit them. Science and technology are presumptively good, but they do not escape a critical examination of their costs and benefits. Progressive support for nanotechnology, while insisting on adequate safety and environmental standards, reflects such a view.

Human Dignity

The term 'human dignity' has recently taken on a new meaning. Conservatives have used it to describe vague concerns about new technologies that force us to reconsider strongly held beliefs. For progressives, 'human dignity' has its original meaning: that of supporting the rights of individuals on the basis of our sense of their shared moral worth as members of the human community (Caulfield and Brownsword 2006). Thus, progressive bioethicists do not insist on one vision of the good life, or impose a single moral belief system on everyone. Rather, they protect and promote the ability of individuals to pursue their own ends, provided they do not impede the ability of others to do the same. Progressive support for patients' autonomy and for access to legal medical treatments and procedures reflects these dignity concerns.

Moral Transparency

For progressives, ethics is not a set of specific, immutable, unchanging laws applied the same way today as a hundred years ago. Rather, our ethics stem from values and beliefs whose expression continues to evolve as they are informed by advances in science, politics, art, culture, and society. New developments cause us to reconsider previously held assumptions, comparing them with our values and, at times, changing our notion of how those values should be expressed. Thus, progressive bioethicists are attuned to these changes, understanding that society may change its opinion on what is right as time passes. We seek to ground bioethical decisions in widely held norms within the community, particularly norms that arise from extensive and informed public debate. The purpose of progressive bioethics is not to impose values, but to help people see how their values can be realized in new contexts within a changing society. Progressive support for *in vitro* fertilization and other reproductive technologies that are broadly desired and understood reflects a belief in public ethics.

Ethical Practicality

As a type of applied ethics, bioethics must remain closely connected to the actual circumstances of our world. Rather than imagine doomsday scenarios of future dystopias or bright utopian futures that may never come to pass, progressive bioethicists must address the questions of the here and now. Of course, preparation can ease the introduction of future technologies into society, but this cannot distract progressive bioethicists from current problems. And we must not simply address these questions theoretically; we must seek to offer practical, realizable solutions. The medical concerns of the developing world, racial and ethnic disparities in access to health care, and the powerful influence of industry on biomedical science and regulation are all issues of great importance that may require messy solutions. Hearkening back to our pragmatist roots, progressive bioethicists must seek to productively address questions in the life sciences in ways that affect the actual world. Progressives' concern with health care as a moral issue, not just an economic one, reflects this concern with practicality.

Toward a Progressive Bioethics

In many ways, there already is a semblance of progressive bioethics. A number of the values described as progressive are among the dominant views of the bioethics academy, including adherence to facts, protection of human dignity, and belief in the potential of science. They are the values that first defined the field, and they continue to hold prominent places in it. But progressive bioethics goes beyond these values. It gives greater prominence to voices that call for a return to concerns with social justice, to the protection of the least among us, and to an engagement with the everyday problems we see around us. It is a call to action, a prompting for bioethicists who share these values to take a more active role in the public and political debate around these issues. Constructively addressing the new moral challenges presented by the life sciences requires an openness to change, an inquiring spirit, and a sense of justice. That is the call of progressivism, one as powerfully inspiring today as it was a century ago.

Works Cited

Berger, Sam. 2007. A challenge to progressives on choice. *The Nation*, July 18.

Borenstein, Seth. 2007. Scientists struggle to define life. Associated Press, August 20.

Bourne, Randolph. 1967. "The failure of Pragmatism." In *American Thought in the Twentieth Century*, ed. D. Van Tassel. Crowell.

Callahan, Dan. 1996. Is justice enough? Ends and means in bioethics. *Hastings Report* 26, no. 6: 8–10.

Caulfield, Timothy, and Roger Brownsword. 2006. Human dignity: A guide to policymaking in the biotechnology era? *Nature Reviews Genetics* 7, January: 72–76.

Cohen, Eric. 2006. The ends of science. *First Things* 167: 26–30.

Cohen, Eric, and William Kristol. 2002. Cloning, stem cells, and beyond. In *The Future Is Now*, ed. W. Kristol and E. Cohen. Rowan and Littlefield. Originally published in *The Weekly Standard* (August 13, 2001).

Dionne, E. J. 1991. *Why Americans Hate Politics*. Simon and Schuster.

Dionne, E. J. 1996a. Back from the dead: Neoprogressivism in the '90s. *American Prospect* 7, no. 28: 24–32.

Dionne, E. J. 1996b. *They Only Look Dead*. Simon and Schuster.

Evans, John. 2002. *Playing God? Human Genetic Engineering and the Rationalization of Public Bioethical Debate*. University of Chicago Press.

Halpin, John, and Ruy Teixeira. 2006. The politics of definition. *The American Prospect*, April 20.

Hinsch, Kathryn. 2005. Bioethics and public policy: Conservative dominance in the current landscape. http://www.womensbioethics.org.

Kass, Leon. 1993. The problem of technology. In *Technology in the Western Political Tradition*, ed. A. Melzer et al. Cornell University Press.

Levin, Yuval. 2006. The moral challenge of modern science. *The New Atlantis* 14: 32–46.

McGerr, Michael. 2003. *A Fierce Discontent*. Free Press.

Moreno, Jonathan. 1994. *Deciding Together: Bioethics and Moral Consensus*. Oxford University Press.

Moreno, Jonathan. 2007. The clone wars. *Democracy* 5: 110–116.

Moreno, Jonathan, and Sam Berger. 2007. "Rush" to judgment. http://www.americanprogress.org.

Moyers, Bill. 2003. This is your story—the progressive story of America. Pass it on. Speech at Take Back America Conference, June 4, Washington.

Mundy, Liza. 2007. *Everything Conceivable*. Knopf.

Schambra, William. 2006. The conservative message machine? Speech at Washington conference titled Building a Better Message Machine: Do Ideas Matter?

2

Can There Be a Progressive Bioethics?

Richard Lempert

Progressive bioethics—the words are not an oxymoron. Far from it; they are more redundant than oppositional. Yet they leave me almost as uneasy, as if they were contradictory. My unease exists because bioethics should be neither progressive nor regressive, neither right wing nor left wing, neither liberal nor conservative. It should be just good, sound ethics applied to the often difficult moral problems posed by present-day medicine and the genomic revolution.

I do not mean to suggest by this that all bioethicists need agree. Respectable ethicists using established modes of ethical analysis have long disagreed on and argued for different conceptions of the ethical across a range of issues far broader than the biologic. Ethical arguments are, however, not all created equal. While some philosophers have gone so far as to argue that ethical discussions are meaningless assertions of preference (Ayer 1936), these discussions in fact influence personal decisions and public policy. Moreover, ethical arguments, as Stephen Toulmin (1950) pointed out, may be well or poorly reasoned, and may be accepted or rejected on this basis.

There is no *a priori* reason to think that sound ethical arguments will necessarily support positions more congenial to liberal political philosophies or to conservative ones. Even progressive bioethics, as I conceive of the term, can lead to conclusions that the right, including the religious right, will find more congenial than the left, for in my view progressive bioethics entails more of a methodological commitment than a commitment to ends. Indeed it is precisely because agenda-driven conclusions are passed off as the necessary implications of ethically driven analysis that a commitment to progressive bioethics is important at this moment in political time.

Let me sketch the ingredients of a progressive bioethics, as I see them. First and foremost is a commitment to sound science and a willingness to build on the best current scientific knowledge to the extent that knowledge is relevant to ethical analysis. Reality should not be distorted so as to favor a desired end, no matter how worthy that end may appear. This is true when facts are known, but it also applies when all science can provide us with is estimates of probabilities. Probabilities should not be distorted or unduly disparaged in the interest of making a more convincing argument.

Consider, as an example, human embryo stem cell research (hESCR) and the clash between progressives who favor federal funding of such research and non-progressives who are opposed. The case for federal funding of hESCR would be ethically stronger and more politically persuasive if it could be claimed that expanding the nation's hESCR investment would soon yield cures for diabetes, multiple sclerosis, and Alzheimer's. It would be similarly strengthened if we were certain that no other source of stem cells could ever adequately substitute for human embryos. Progressive bioethicists cannot, however, make these claims. The claim that hESCR will soon lead to cures for dread diseases is today an expression more of hope than of fact, and the jury still appears out on whether the versatility and adaptive potential of embryonic stem cells can be matched by stem cells derived from other sources (Weiss 2007).

But those opposed to hESCR face similar strictures. Their arguments would be philosophically and politically more persuasive if they could honestly argue that the beneficent potential of stem cell research was a mere chimera, or that we could be confident that fetal cord, adult skin, or other stem cells had as much potential for realizing disease cures as human embryonic stem cells. They could also make a stronger case if they could honestly claim that progress in deriving cures through stem cell research was not likely to be delayed by focusing federal funding on stem cells from sources other than human embryos, and if they could honestly argue that the choice of what kinds of stem cell research to fund was based entirely on scientific rather than political reasons. But these claims, like proponents' suggestions that cures for dread diseases are imminent, cannot be honestly made and so should not be made at all. The cases for and against federal funding of hESCR should proceed on the best current scientific estimates of the potential value of such research and its alternatives, rather than on disputants' wishes about what the truth will turn out to be.

Placing science first means there is no guarantee that bioethical analysis will inevitably favor the conclusions associated with established political progressive positions. Consider, for example, progressive support for allowing women to make choices about childbearing free from state coercion. The political case for largely unfettered freedom, and perhaps even the moral case, has been weakened by medical progress in lowering the gestational age at which fetuses can survive as infants and in diminishing, for given birth weights, the likely lifelong harms associated with premature delivery. Should further research reveal that past a certain gestational stage the fetus feels pain from abortion procedures, both the political and the ethical case for state regulation will be strengthened.

Neither the reduction in the time to onset of fetal viability nor (should it be shown) the ability of the fetus at a certain gestational stage to feel pain is a philosophical trump that defeats that ethical case for a woman's right to choose whether she wishes to carry a baby to term. An ethicist could still argue that a woman's right to control her own body and her role in reproduction free of state interference is a higher value than any interest of the fetus, even an interest in avoiding pain, and it could similarly be argued that access to abortion is ethically justified or even mandated when pregnancy results from rape or incest or would endanger a woman's health. But our capacity to preserve life at younger gestational ages enhances ethical arguments made by opponents of late-term abortion and is likely to aid political and judicial attacks on abortion access even with respect to earlier-term procedures. Such effects are likely to be multiplied if science finds strong evidence that abortion causes the fetus to feel pain. A progressive bioethics can seek to answer the arguments that such newly found facts would support, but if the facts are rooted in sound science, progressives should neither ignore nor deny them. Indeed, one can imagine facts that would lead at least some progressives to rethink their ethical positions.

A second commitment that should characterize progressive bioethics is a commitment to reasoning from principles. Some of these principles relate substantively to progressive positions; others concern analytic methodology and are largely content free. These latter include, most notably, the many flavors of utilitarian, deontological, and virtue-ethics arguments on which philosophers regularly draw in ethical analyses to justify the conclusions they reach. The progressive approach does not necessarily mandate a choice among these modes of justification, but it does preclude some approaches found in bioethical commentary.

One excluded approach turns to religious positions based on faith for answers to ethical questions, leaving no additional role for reason. This exclusion extends to arguments that seem to reason from ethical principles toward conclusions but in fact collapse if the faith-based premises on which they rest are denied. Arguments of this sort are not bioethical arguments and have only second-order relevance (explained below) to bioethics debates. Rather, they draw their propositions from religious dogma in order to question the morality of medical or other bioscience practice or otherwise to answer bioethical questions. Opposition to hESCR grounded directly or derivatively in the proposition that destroying human embryos is a sin rests, for example, on a religious argument, not a bioethical one. Resting bioethical conclusions on nothing more than religious dogma or related perceptions of "God's word" is inconsistent with the progressive view of how bioethical argument should proceed.

It is not, however, illegitimate or wrong to draw implications from religious ethics for biological practices so long as what is done is not disguised as something else, and progressives should recognize this. The view that the soul arrives to create a fully human life the minute an egg and a sperm come together, and that therefore the destruction of an embryo is the destruction of human life, akin to killing a baby, may not be widely shared, but it cannot be proved true or false. It just is. Moreover, it is not an evil view (like the view that people of some race are inherently less worthy than others), and it deserves to be respected. Respect does not mean accepting the prescriptions that follow from the view, nor does it require treating the belief as an acceptable bioethical argument, but it is also not an empty commitment.

One reason for respect is that although religious mandates should not be confused with bioethical arguments, they may give rise to arguments that progressive bioethicists should answer, such as the argument that a person's integrity is offended when, as a taxpayer, he is forced to help fund research he considers deeply immoral. Even if respect for the views of others does not entail supporting morality-based tax avoidance,[1] it may, as an ethical matter, support lesser accommodations. The implications of people's views of right and wrong, whatever their roots, are a proper concern of bioethicists. Indeed, the degree to which moral preferences are shared is a fact, one that is even subject to loose determination through polling. If large numbers of people are troubled by an action, utilitarian philosophies, at least, would argue that their reactions have implications for moral assessment. The role of personal moral preferences is, however, of a second order; the fact that some people are troubled by a procedure or action is not a (first-order) bioethical argument for the immorality of that procedure. Their concern does mean that even if the principal action is ethical, there are utilitarian reasons for not engaging in it, because to do so would cause people pain.

Consider again the stem cell example. There are strong bioethical reasons for advocating the pursuit of hESCR. These have to do with the virtues of curing disease and the benefits (happiness) this will bring people. Deontological arguments can also be made to the same end. Indeed, it is hard to identify ethical arguments against engaging in hESCR, apart from the faith-based argument that, despite their early life stage, the embryos used in hESCR are, from a moral standpoint, human, and thereby entitled to the same respect accorded all humans even if, as is the case, they are doomed. The second-order argument that a relatively large number of people will be offended, saddened, or otherwise distressed should hESCR proceed cannot counter the bioethical case for advancing cures, because it does not even address it.[2] Suppose, however, that it became possible to derive stem cells from adult tissues that were identical to embryo stem cells in all therapeutically relevant respects. Then the existence of people distressed by the prospect of killing embryos should influence the ethical debate, and respect for those distressed would make the moral case for research using embryo-derived stem cells crumble since the health imperative would no longer require using embryonic stem cells.

A second approach that progressive bioethics excludes is the "yuck factor" test—what Kass (1997) terms the wisdom of repugnance. In such a test, an individual or a group is doing nothing more than asserting personal preferences as natural. To assert that something is disgusting is not to make an ethical argument, for it advances no reason for judgment apart from personal preference. Claims based on a person's disgust fail as ethical arguments because they cannot be countered by reason. This does not mean that a person's sense of the disgusting cannot motivate bioethical argument, for emotional reactions often stimulate reasoned claims.

One must also distinguish a shared sense of "yuck" from the assertion of a personal preference. Progressive bioethics need not deny, and indeed itself asserts, the validity of moral intuitions. The sense that some behavior is disgusting may be rooted in an act's immorality (e.g. disgust at torture), and a consensus that an act is immoral may be shaped by a shared gut-level sense that it is disgusting. Nevertheless, resting bioethical analysis and policy recommendations on what is collectively disgusting is problematic. The first difficulty is getting the collective reaction right. Those who find an action disgusting may feel that any morally sensitive person would feel the same way, but this does not make it so. Nothing is quite as hollow as an appeal to how "everyone feels" when everyone does not feel the same way. Also, the existence of a strong shared reaction to an act does not necessarily suggest a moral truth (Harris 1998; Nussbaum 2004). The pattern of death sentencing in the United States suggests, for example, that people are more repulsed by interracial murders when whites are the victims than when blacks are killed, but the fact that this sentiment seems widely shared does not mean that murder of a white by a black is morally more deplorable than the similar killing of a black by a white. Similarly, the moral case for hESCR must rest on more than the fact that a few people are instinctively repelled by descriptions of how stem cells are extracted from embryos or the procedures that egg donation requires.

It is particularly hard to build a moral case on "yuck" reactions in the medical sphere, for many well-accepted medical procedures—colostomies, for example—trigger emotions of disgust in ordinary people, and few people, if any, question the ethics of employing them. More generally, the assertion that visceral reactions are evidence of fundamental ethical principles must be made with considerable humility, for history tells us that both visceral reactions and intuitions about what is ethical change. *In vitro* fertilization, for example, was once, and still is by some (see e.g. May 2003), opposed on the ground that it is unnatural procreation and that intuitively it seems wrong. But couples use *in vitro* technologies to start the baby-growing process every day. The virtues of *in vitro* technologies are manifested in the children produced, and opposition on the basis of moral intuition has largely disappeared, apart from some religious forums, in countries where these procedures are common.

Thus, even if collective disgust, like shared religious views, may be a second-order consideration in bioethical analysis, progressive bioethicists should seldom if ever rely on revulsion in grounding their arguments. Bioethics will be advanced only by examining and identifying the moral status of those aspects of a situation that lead to widespread disgust. If these aspects cannot be identified, we have no first-order reason for preferring one group's disgust to another group's lack of qualms.

Progressive bioethics is not only about taking science seriously and having commitment to reasoned analysis as a methodology. As importantly, it involves reasoning from a set of foundational principles. Although these principles may themselves be philosophically justified, progressive bioethics often treats them as taken-for-granted starting points. One such principle is that there is a positive value to the protection and promotion of human life, people's health, and the quality of people's lives, including freedom from pain. This means that actions that protect human life, work to counter disease and debilitation, and help people avoid pain are presumptively good from a progressive bioethics perspective.

Progressives can and do differ about whether human life has the same ethical relevance at all its stages and, indeed, when it can be said that life that counts as human begins and ends. Most progressive bioethicists would argue that, whatever the human-life interests of embryos and fetuses, they should count less heavily than the human-life interests of those who have been born. Some would further argue that at certain stages (the early embryonic stage, for example) there is no human-life interest that merits protection. Similarly, at the other end of the life span, most progressive bioethicists would agree that life has ended when a person's brain has stopped functioning and there is no prospect of recovery, even if the heart is still pumping and the body is still warm. Thus, keeping a body vital after brain death to facilitate transplants that will save lives is not only unproblematic to progressive bioethicists, but it is also likely to be the morally right course of action.

In addition, few progressives would argue that the principles of promoting life, health, and quality of life are lexically ordered, such that life is always more precious than health or the quality of the life lived. Apart from the value of autonomy (discussed below), progressive bioethicists are likely to support rights to treatments that risk death to stop intolerable pain or the right to chose riskier interventions over less risky ones if the riskier intervention promises to leave the patient functioning at a noticeably higher level than could be expected after a less risky (i.e. less likely to be fatal) procedure.

Progressive bioethics also eschews romanticism in assessing the relationship between values. A progressive bioethicist might believe that human life begins with conception and accept the fact that a newly formed embryo is human life. But the progressive ethicist would also recognize that most laboratory-created human embryos will never be implanted in a womb, perhaps to grow to infancy. Rather, "excess" embryos are likely to be stored for a period and then discarded. Recognizing this yields a bioethical analysis that applies to our world rather than to an imagined world in which all embryos would, in the end, be children.

THESE ARE ENDS, CONTRA p. 23

Similarly, when it comes to laws surrounding abortion, progressives are concerned not just with a woman's right to choose but also with whether outlawing abortion would prevent fetal deaths or simply drive abortions underground, saving few fetuses but greatly increasing the danger to the women who carry them. To the extent that the latter is likely, the progressive bioethicist, even if sympathetic to right-to-life concerns, sees outlawing abortions as hard to justify. Bioethics is, after all, concerned with what is justified in this world, and not about religious preparation for the next one.

Although the fetus's life and potential for personhood matter to most progressives, the value of the mother's life and health, her right to make autonomous decisions, and the life the fetus would enjoy if brought to term also matter. If, for example, a fetus is known to have a genetic defect that will result in severe retardation and an early painful death, it would be the rare progressive ethicist who would find an abortion immoral even if the mother faced no danger from carrying the fetus to term and the diagnosis came rather late in the fetal developmental cycle. Indeed, some bioethicists would see early termination of the pregnancy as the ethically superior action.

Progressive bioethics also has little use for symbolic statements that serve primarily to assert the moral supremacy of one group's values over those of another group. It questions efforts to write sectarian ethics codes into law, and it is particularly suspicious of such efforts when a behavior to be prohibited has long existed and long been ignored. When the legalization of contested virtues carries costs, such as moving abortions from the clinic to the back alley or scaring doctors away from the best available pain control, progressives are likely to regard laws as morally flawed, even if they would regard the enactment as morally justified if its enactment and enforcement imposed no or different costs.

Progressives are not libertarians. Few, if any, would dispute the moral propriety of discouraging drug addiction and preventing the diversion of pain killers to black markets. But in assessing the moral status of enforcement guidelines designed to prevent the abuse of addictive painkillers, progressives would weigh in the moral balance interference with the doctor-patient relationship and possible chilling effects on effective pain-management therapies.

A second bedrock progressive value is autonomy. Progressive bioethics incorporates a presumption that those most directly affected by medical and related decisions have a right to make them. A corollary is that a person has the right to access the information needed to make an informed choice and should be free from coercion in making this choice. Autonomy also incorporates a right to privacy and the ability to control access to information about one's health. These values inform progressive bioethics on a wide variety of issues, including the doctor-patient relationship; the obligation of hospitals and caretakers to honor advance directives; rules governing drug trials; the circulation of patient information to relatives, insurance companies, and other third parties; and the availability of embryos for stem cell research.

Difficulties arise in several situations. One is when the person most directly affected by a medical decision is not capable of informed autonomous action, as is often true of young children, senile adults, and some people suffering mental disabilities. Difficulties also arise where honoring one person's autonomy or right to privacy will adversely affect others. A woman found to have a breast cancer gene might, for example, be unwilling to tell her sister what she has learned, though if the sister knew she would be better able to assess her own cancer danger and protect herself from it. A different difficulty arises when patients, asserting their autonomy, seek insured or subsidized payment for a treatment that is not cost justified, thereby raising the cost of necessary health care for everyone. Not surprisingly, the label 'progressive' says little about how a bioethicist will resolve these or other hard problems.

The elephant in this room, of course, is abortion. A woman cannot have an abortion without killing a fetus, so the autonomous choice for an abortion is arguably at the expense of another human life. One way to deal with this suggested conflict, which we see in some ethical analyses, is to deny the personhood of the fetus or to give it less than fully human status. Either allows the construction of an ethical case for a woman's right to choose when her life or health would be harmed by carrying the fetus to term. Denying personhood entirely frees the woman from even these constraints, but acknowledging that the fetus can to some degree claim the moral status of a person raises difficulties for those who would justify abortion whenever a woman prefers not to give birth.

One way some address this situation while retaining a woman's right to choose is to echo Justice Harry Blackmun's approach in *Roe v. Wade* and treat a fetus as differentially human depending on gestational age. For those taking this stance, the fetus during the first trimester is regarded as essentially lacking personhood, and a woman's choice to abort presents no ethical problems. During the third trimester, when the fetus has increasingly taken on the physical characteristics of infancy with increasing potential to survive outside the womb, the ethical balance shifts and a very good reason, such as a serious threat to the woman's health, is needed to justify an abortion. During the second trimester, matters are more ambiguous, and shift as fetal development proceeds. Here the likelihood that a woman would secure an abortion regardless of what the law or abstract ethical analysis proscribes might play a role in practical ethical assessment.

A different approach to this conflict avoids contending with the possible personhood of the fetus. It argues that, no matter what the fetus's status as a person, ethics does not allow a regime that enlists a woman against her will and at some cost (even if only the normal stresses of pregnancy) to devote her body to the nurturance of another person (Thomson 1971). This view is consistent with the law's reluctance to require people to come to the aid of others even if the other's life is in danger and the rescuer would be barely inconvenienced (Regan 1979). For some this is no answer, because they regard the law's refusal to insist on altruistic behavior as a moral shortcoming. Moreover, opponents of this view point out that legal obligations to aid others do exist in special relationships, such as the parent-child relationship or after one has begin to extend aid. The fetus-host relationship is not a parent-child relationship, but in allowing a fetus to develop beyond the first few months of pregnancy a woman has arguably begun to extend the kind of aid that can impose an obligation to perfect a rescue. Thus, even if the legal analogy were accepted as a moral benchmark, some would still argue that gestational stage is relevant to the ethics of abortion.

A third bedrock principle of progressive bioethics is valuing equality. Progressives believe that morality requires that people be treated more or less equally, particularly with respect to fundamental goods such as health care. Progressives are wary of distinctions in treatment, research, and the availability of health services that disadvantage the less powerful, particularly minorities and women. For the progressive bioethicist, some of the most egregious examples of unethical behavior have been of this sort. The Tuskegee syphilis study, which, in the interest of understanding how syphilis progresses, allowed its black participants to go untreated for more than two decades after an effective treatment was available, is an iconic example (Jones 1993). Indeed, as commonly understood, the decision to allow study participants to remain untreated after penicillin became the standard syphilis treatment was unethical to the point where it should have been criminal. This is not just because people who could have been helped through drugs were allowed to suffer and die, but also because it appears inconceivable that the study would have continued to the point it did had the men in the study been white rather than black. Although it has not achieved the same iconic status, the delayed attention to AIDS because it was perceived as a "gay disease" is similarly deplorable from a progressive ethical perspective.

Living in this world, progressives recognize that income differences will inevitably affect the quality of medical attention people receive, but they believe ameliorating income-based discrepancies to the extent possible and erasing racial and gender differences in care quality are ethical mandates. They similarly object on ethical grounds to research funding that, in relation to the harm caused, disproportionately invests in conditions that are especially likely to afflict better-off groups, including the wealthy, whites, and males.

The last core principle I shall mention is justice. Although related to equality, it is not the same thing. Justice involves fair treatment, which is not the same as similar treatment. For example, a kidney donor whose remaining kidney later fails might have a justice claim to move to the head of the queue for kidney transplants, but would not have an equality claim. Perhaps nowhere are justice claims more prominent than in discussions of the obligations owed by nations and companies to individuals in the developing world who participate in drug trials that would be more difficult if not impossible to arrange in developed nations. It has been argued that justice is offended when people instrumental in the development of a drug are, for financial reasons, unable to benefit from the advances they have made possible, and that, at a minimum, drug companies have an obligation to continue treating drug trial participants in the developing world for as long as their health requires after the trial has ended (Glantz et al. 1998).

Justice, as well as equality, is important when analyzing other issues, such as obligations to invest in drug development for so-called orphan diseases, similar obligations to invest in treatments for diseases largely confined to the developing world, and the obligation to furnish or (for drugs under patent) allow the furnishing of low-cost medications to people who cannot afford to pay normally charged prices.

Two things should be obvious from this list. One is that progressive principles do not yield determinate results, even if some arguments are ruled out and some principles enjoy a privileged status. Conflicts can exist between core progressive principles, and largely like-minded people can differ in evaluating the priority of principles and determining what these imply. For example, justice and equality considerations, together with the progressive commitment to advancing health, counsel the provision of effective drugs at low cost to those who require them, even if this means overriding patent protections to produce and allow the distribution of generic equivalents. The concern for improving health similarly supports efforts to stimulate the production of effective new drugs. And an intellectual-property regime that allows drug patents to be ignored in countries where drugs are beyond the means of most disease sufferers reduces private incentives to invest in drug development, particularly for drugs that treat diseases endemic to poorer countries. So what does progressive bioethics say about allowing developing nations to produce or secure generic versions of life-saving drugs that are still under patent? Not surprisingly, bioethicists, even those who call themselves progressives, do not speak with a single voice.[3]

What those who call themselves progressives should hold in common is a commitment to base assessments not on speculation as to how the world operates but on the best available relevant information. Thus, the progressive bioethicist should be skeptical of the easy assumptions that, on the one hand, undercutting patent protections to benefit the ill in the developing world will dramatically reduce investments in new drug research and, on the other hand, that, because most people in the developing world cannot afford expensive drugs, allowing generics to be freely marketed in developing countries will have no effect on a patent holder's bottom line. Good empirical answers do not, however, exist for all the questions that bioethicists might ask about human and organizational behavior, and the real-world policies advocated in bioethical analyses, including those of progressives, must rest on empirical assumptions that are less than fully supported. Where this is the case, it is likely that policy and other preferences will influence the facts assumed, even by people making a good-faith effort to remain "objective."

The second point that should be obvious is that the concerns and methods of progressive bioethics do not differ that much from the concerns and methods of most bioethicists. Even if one associates progressives with liberals and sees the opposite of progressive as conservative, there is no good reason why on most issues there should be large, consistent differences. Many of the matters that concern bioethics, like the conditions under which informed consent should be required and what adequate consent entails or the obligation of lab scientists to share materials developed with the aid of federal funding, yield answers built on principles widely shared in liberal and conservative communities. Other questions, like that of a genetic counselor's obligations when a gene test incidentally reveals that a woman's father could not be the person she calls "Dad," are so removed from politics that their answers cannot be placed on a liberal-conservative spectrum. Still other questions are difficult no matter one's political ideology. As a result, there are issues on which one finds disagreement among ethicists who have generally similar political values and issues on which one finds agreement among people of different political persuasions.

The situation is different with regard to religiously driven ethicists, who seek to import their religious beliefs into bioethics principles, arguing, for example, against behaviors they regard as sinful. The methodologies used by the progressive bioethicist and the religious ethicist are different, even if the latter will also muster rational, secular arguments to support their positions. Conclusions reached are often different as well. Progressives can, for example, find no principle in bioethics that justifies denying interested couples contraception or precluding abortions when a mother's life is in danger. Some religiously driven ethicists find reasons to support both prohibitions.

However, even across this divide it is easy to overstate differences, at least with respect to outcome preferences. The world has many religions, and the tenets of some accord with progressive principles. Moreover, even within religions whose dogmas apparently conflict with progressive preferences, there will be faithful members who share progressive views, often justifying them by aspects of their religious dogma that church leadership has ignored. Moreover, as with liberal and conservative differences, many bioethics issues do not line up on a spectrum likely to differentiate those who assess behavior by reference to religious values from those who reject the idea that religion can provide first principles.

Should we then reject the idea of a progressive bioethics altogether and go back to a world populated by bioethicists without any adjectives?

Part of me likes this idea, eschewing labels for a group whose members think differently about many issues. Yet there are senses in which progressive bioethicists can be distinguished within the bioethics communities, and ultimately they justify a separate label.[4] Progressive bioethicists tend to justify and advocate positions that appeal to members of the liberal and other progressive political communities. The route to these preferences should be from ethical foundation to political preferences, rather than the reverse, but it can sometimes be difficult to separate the two. If justification through "yuck" reactions or other gut instincts is not an acceptable way of arguing for an ethical position, a moral intuition that lies behind such reactions is not delegitimized on that account. Rather, the reaction is cause to examine first principles and to search for reasoned justifications for the intuition—justifications that go beyond the claim "I and others like me feel this way."

Progressives tend to share a number of moral intuitions. They become offended when people are treated worse than others with respect to health care, including less research attention to certain diseases, because of their race, gender, sexual orientation, ignorance, or poverty. They do not like it when people try to impose principles drawn from religious doctrines on those with different beliefs, or when groups seek to regulate private behavior (especially voluntary sexual behavior) that puts no one external to consenting participants at significant risk. Progressives are also offended by dishonest arguments made in support of preferred beliefs, such as arguments regarding the effects of abortions on women experiencing them that misstate social science findings to make the exceptional reaction sound usual. They also disapprove of, though some cannot help but admire, the way conservative advocates manipulate language to support their positions, such as the transformation of the fetus into an *unborn child* and the creation of the term *partial-birth abortion*. The problem progressives have with such nomenclatures is that they are designed and employed to promote political decisions based on emotion, rather than to encourage clear thinking about difficult ethical questions like those that abortion raises. Progressives also find the use of such language troubling because it seems to work, and if they could find their own emotional keys to political persuasion many progressives would, I expect, employ them despite their abstract belief in reason.

There are two issues that are almost iconic in their capacity to distinguish progressive bioethicists from those whom they see as principal antagonists: abortion and hESCR. Recognizing that there are difficult issues surrounding abortion and that the fetus, while not a baby, is also not an inanimate object with no relation to human life, a progressive weighing of conflicting values supports a woman's right to choose whether she wishes to carry a baby to term, at least when the decision is made within the first three to six months of pregnancy or at any time if a woman's life is endangered. Also, while progressives can understand and respect opposition to hESCR rooted in religious doctrine, from a bioethical standpoint allowing such research seems almost a "no-brainer." It is true that human tissue is involved, but in all other respects no one except a minority of the religiously motivated would call the embryos from which stem cells are extracted human beings.

Apart from their DNA, embryos have none of the features that make a person, or even a fetus, distinctly human. No one could tell from looking whether a particular embryo was human in origin, and there can be no pain associated with hESCR because embryos have neither nervous systems nor brains. Moreover, the embryos whose stem cells would be used in research are not produced for research purposes but are a by-product of people who seek to bear children with the aid of *in vitro* conception. There is no chance that the embryos used for research will become even "unborn children," for these embryos would otherwise be destroyed by being discarded. "Embryo adoption" is almost unheard of, and everyone interested in embryo adoption could do so without affecting research.

On the benefit side, hESCR promises to enable treatments of diseases and other conditions that could save lives or improve their quality in ways that no other currently contemplated therapy might. While unforeseen obstacles may mean that this promise is never realized, the only way to discover this is through research. It is also possible that umbilical cord stem cells, bone marrow stem cells, or even adult skin stem cells could do the work that human embryonic stem cells are expected to do, but we don't know this for a fact, and there is reason to doubt whether cells from these sources would be as versatile as those from embryos. Moreover, even if these other possibilities would pan out, waiting for them to come to pass would delay, perhaps by decades, advances that might come from working today with embryonic stem cells.

The George W. Bush administration's compromise on stem cell research, which allowed federal funding only for the few cell lines that pre-existed President Bush's August 9, 2001 decision on the issue, is from a progressive bioethics standpoint particularly perverse. To begin with, it was built on bad science, because it assumed the preexisting stem cell lines were sufficient to allow a rich array of research to proceed when it was clear they were not (Association of American Universities 2005). It was also inconsistent, in that if it were truly immoral to use human embryonic stem cells for research purposes, the private sector should not be allowed to freely do so, and if it is not immoral, the federal government should be supporting hESCR because of the moral imperative to work to cure disease. It was also counterproductive. It discouraged some leading biological researchers from working in American laboratories. It imposed unnecessary costs on universities, their faculties, and their students because those universities that sought to support this research were forced to duplicate existing facilities and to ensure that no expenditures that might in the smallest degree be traced to federal funds, including the universities' general indirect cost pool, went to support research with embryonic stem cells (Moreno et al. 2007). Finally, to the extent that aspects of this research should be regulated, by not funding such research the federal government abandoned its regulatory authority. It is no wonder progressives' hackles were raised. Their question is, why weren't everyone's?

Much of my description of how progressive bioethicists view, analyze, and resolve contemporary issues in bioethics is familiar in the academy, for most academic bioethicists acknowledge as legitimate, and in large measure share, both the methodological commitments and the values I attribute to progressive bioethicists. Thus, "academic bioethicists" might be substituted for "progressive bioethicists" at many places in this essay, and, with the caveat that there is greater diversity in the academy in general than among self-identified progressives in particular, most of what I write characterizes the approaches and views that dominate in the academy. But there is one way in which progressive bioethics attempts to go beyond the mere academic: it seeks to actively engage the policy, and hence political, world. Progressive bioethics speaks not just to other ethicists and to bioethics students, but also is a distinct voice in the political marketplace of ideas. Its arguments are, or should be, constructed to appeal not just to intellectual elites, but to political leaders and ordinary people—to blacks and whites, churchgoers and non-churchgoers, white-collar workers and blue-collar workers, the working woman and the full-time homemaker, the employed and the unemployed, college graduates and high school dropouts. These are the people who determine how bioethical arguments come to be embodied in state practice. They ultimately determine what restrictions will be placed on abortions or the degree to which hESCR will be federally promoted. If these people are not reached, the best bioethical analysis can be an empty exercise, satisfying for like-minded people to contemplate but ultimately ineffectual.

There are significant challenges, which progressive bioethicists have not yet mastered, to being effective voices in the public sphere. One challenge is to balance the demands of honesty and principle with the demands of effective persuasion. In the battle over abortion rights, for example, and in particular the battle over Medicaid-funded abortions, the most effective popular argument might, with some empirical support, be encapsulated in the slogan "unwanted children kill wanted children" (Donahue and Levitt 2001). But progressive bioethicists have not made this argument and should not. For one, we don't know if it is true, as research suggesting it might be has been strongly criticized (Joyce 2004). But more important, it would be very easy for a campaign based on such a theme to feed, and feed on, racism, with many people thinking in terms of unwanted black children and wanted white ones. Hence, to engage the public with this argument would be to subordinate important progressive principles to the attainment of a desired end.

There are, however, ways that progressives can frame bioethical issues to make their points more persuasive and their conclusions more generally acceptable. Perhaps the greatest triumph of the anti-abortion movement has been to transform the fetus into the *unborn child*. Arguments that are easy to ignore when one is talking of fetuses resonate when the conversation is about unborn children. Yet even media and reporters sympathetic to abortion rights now often use this term. This transformation has worked for two reasons. One is that abortion opponents use it consistently in almost all they say and write. Anyone who wishes to quote what they say must use it too. But the other reason is that it contains a good deal of truth. Long ago, some opponents of abortion rights referred to fetuses as *babies*. This never caught on, because fetuses and babies are separate concepts, and no matter how much a fetus may come to resemble a baby, there remains a hard-to-defeat intuitive difference. But a fetus is a potential child, with the potentiality indicated by the word 'unborn'. One cannot dispute the fact that a fetus carried to term and born alive would be a child.

Progressives need to employ similar tactics when it comes to influencing how issues are framed. They have certainly tried, and to some extent have succeeded, in the abortion context by emphasizing that an abortion involves a woman's right to *choose*. But they are at a disadvantage because those who would restrict abortion and who accuse abortion-rights supporters of killing unborn children not only can use more emotionally laden terms than the defenders of abortion, but can use visuals, both appealing and shocking, to make their case that abortions kill soon-to-be-children.

When it comes to hESCR, however, the advantage is reversed. Here progressives can proceed on two fronts. One, already used effectively, is to show the human face of diseases that might be cured if stem cell research is allowed to proceed. We might think of this as the Michael J. Fox or Nancy Reagan effect. The second, which we have not yet seen, is the creation of language that emphasizes an important morally relevant fact: that the embryos whose stem cells would be extracted for research would be destroyed in any event. Rather than speak of *human embryo* stem cell research, progressives seeking to win the battle of public opinion would be wise to talk about *discarded embryo* stem cell research. This has the virtue of accuracy, and does not call to mind the image of a potential child in the way the word 'human' does.

I still am made uneasy by the term 'progressive bioethics', because I believe that what so-called progressive bioethicists stand for is, for the most part, nothing more than sound bioethical analysis. However, I ultimately accept the idea that there is a place for the term and for the movement it represents. Progressives accent values in the bioethical marketplace of ideas somewhat differently than people without progressive commitments, and the progressive interest in practical policy can justify the use of a political term to describe what is largely an academic movement. Indeed, the major challenge confronting the progressive bioethical movement is neither clarifying its moral bases nor finding consensus on progressive principles. Rather, it is to develop the political effectiveness needed to advance the application of sound bioethical principles in public life.

Notes

1. The practical answer to the objection has been provided again and again; the taxpayer has no standing to complain. The religious objector to hESCR stands in the same position as the religious objector to war. A respectable ethical argument may be made against coercing through taxes contributions to actions that offend, but it is an argument that in the United States has never carried the political day.

2. Of course, if enough people felt this way, principles of democracy might allow the majority to forbid federal funding for hESCR or even all hESCR no matter what the funding source. However, this kind of determination would not be based on bioethical grounds, but on religious or other ethical grounds, coupled with a political system that in most spheres allows a majority to impose its will, whatever its basis, on a minority.

3. Paragraph 30 of the World Medical Association's Declaration of Helsinki which deals with ethical issues in medical research involving human subjects provides that "at the conclusion of the study, every patient entered into the study should be assured of access to the best proven prophylactic, diagnostic and therapeutic methods identified by the study." Efforts to amend this paragraph to deal with matters left open (e.g. for how long must access be provided) have failed because of an inability to reach consensus. The most that could be agreed upon at the 2004 meetings was that a description of the plans for providing post-trial access arrangements or other care must be part of the protocol that is initially submitted for ethical review.

4. Whether 'progressive' is the best way to label this group is, for me, an open question, but at least for the moment it seems to be the label this group has.

Works Cited

Association of American Universities. 2005. Letter to all members of the House of Representatives regarding the Stem Cell Research Enhancement Act of 2005, May 23.

Ayer, Alfred J. 1936. *Language, Truth, and Logic*. Gollancz.

Donohue, John J., and Steven D. Levitt. 2001. The impact of legalized abortion on crime. *Quarterly Journal of Economics* 116: 379–420.

Glantz, Leonard H., George J. Annas, Michael A. Grodin, and Wendy K. Mariner. 1998. Research in developing countries: Taking "benefit" seriously. *Hastings Center Report* 28, no. 6: 38–42.

Harris, John. 1998. *Clones, Genes, and Immortality: Ethics and the Genetic Revolution*. Oxford University Press.

Jones, James H. 1993. *Bad Blood: The Tuskegee Syphilis Experiment*. Free Press.

Joyce, Ted. 2004. Did legalized abortion lower crime? *Journal of Human Resources* 39: 1–28.

Kass, Leon R. 1997. The wisdom of repugnance. *New Republic*, June 2.

May, William E. 2003. Begetting vs. making babies. In *Human Dignity and Reproductive Technology*, ed. N. Lund-Molfese and M. Kelly. University Press of America.

Moreno, Jonathan, Sam Berger, and Alix Rogers. 2007. Divided we fail: The need for national stem cell funding. Center for American Progress, April 12.

Nussbaum, Martha C. 2004. Danger to human dignity: The revival of disgust and shame in the law. *Chronicle of Higher Education* 50, no. 48: B6.

Regan, Donald H. 1979. Rewriting *Roe v. Wade*. *Michigan Law Review* 50: 1569–1646.

Thomson, Judith J. 1971. A defense of abortion. *Philosophy and Public Affairs* 1: 47–66.

Toulmin, Stephen. 1950. *An Examination of the Place of Reason in Ethics*. Cambridge University Press.

Weiss, Rick. 2007. Scientists use skin to create stem cells. *Washington Post*, June 6.

II

Bioethics as Progressive

3

Politics, Progressivism, and Bioethics

R. Alta Charo

Within three months of the announcements from Wisconsin, Japan, and Massachusetts of their respective successes at deriving induced pluripotent stem cells (iPS cells) from adult somatic tissue (Lowry et al. 2008; Nakagawa et al. 2008; Park et al. 2008; Takahashi and Yamanaka 2006; Yu et al. 2007), thus avoiding the need to work with human embryos, a leading conservative bioethicist was reprising his earlier call for a ban on the better-known technique of somatic cell nuclear transfer (SCNT), also known as "research cloning." Grounding his reasoning in the assumption that science which offends public sensibilities ought to be banned unless manifestly necessary for a competing and important medical need, he wrote:

Cloning for the purpose of biomedical research has lost its chief scientific raison d'être. ... The time is also ripe for a separate bill [calling for] a (four- or five-year) moratorium on all de novo creation—by whatever means—of human embryos for use in research. This would block the creation of embryos for research not only by cloning (or SCNT), ... but also by IVF. ... The new iPSC research ... suggests that our society can medically afford, at least for the time being, to put aside further creation of new human life merely to serve as a natural resource and research tool. *We can now prudently shift the burden of proof to those who say such exploitative and destructive practices are absolutely necessary to seek cures for disease, and we can require more than vague promises and strident claims as grounds for overturning the moratorium.*" (Kass 2008, emphasis added)

The former chair of President George W. Bush's Council on Bioethics was not alone in asserting that, with the likelihood that iPS cells could someday (assuming problems with oncogenes and viral vectors were overcome) substitute for human embryonic stem cells, not only should federal funding be withheld from hES cell research, but the research itself should be wholly forbidden (Krauthammer 2007; Anderson 2007).

Shifting the burden of proof so that a presumption of freedom to do research is replaced with a presumption that research may be banned even in the absence of concrete harms to actual persons is a radical notion, and yet it has been implemented in Arkansas, Indiana, Michigan, North Dakota, and South Dakota. Science, it would seem, has now become thoroughly enmeshed in long-standing constitutional and political philosophy debates concerning the authority and wisdom of governmental morals regulation (Hunt 1999; Eckenwiler and Cohn 2007).

Whether the proper role of government is to ban research that is socially disruptive or to accept the serendipitous changes it may bring is a question of political philosophy that is at the heart of many debates concerning public policy responses to bioethics dilemmas. More than the specifics of a technology or its applications, it is the presence or absence of a presumption of freedom of inquiry, of action, and of personal identity that can determine whether the public policy response should be to ban a technology or merely to regulate it in a manner that guards against concrete harms to actual persons.

Back in the late 1990s, when cloning was about Dolly and not about stem cells, representatives of national bioethics commissions from around the world had gathered at an ancillary meeting during the International Association for Bioethics's annual conference to discuss the possibility of a global consensus position on reproductive cloning. The conversation moved along predictable lines, with all agreeing that reproductive cloning was unsafe for now, and that even if it ever was safe, it was generally a distasteful idea. The reasons for finding it distasteful ranged from concerns about excessive parental egoism to disregard for children's expectation of uniqueness to a (mathematically unpersuasive) diminution in human genetic variation. Countering these arguments were predictable bows in the direction of personal autonomy in reproductive decision making.

Although consensus was easily reached on the merits (or lack thereof) of reproductive cloning, proposals to harmonize national policies were more troublesome. While many countries had invested unfettered power in their national governments, the United States was faced with questions about federal versus state jurisdiction and about constitutional protections for reproductive choices. In addition, although many countries had political freedom to craft compromise positions in the realm of human reproduction, U.S. politics was strongly influenced by the bitterly divisive abortion debate, a debate that had led pro-choice groups into extreme libertarian positions lest the principle of reproductive choice be undermined.

The conversation revealed that any effort to harmonize national policies would depend less on a consensus concerning the technology and more on whether national legal and political cultures could be harmonized. And further, if there is a genuine value in the diversity of national cultures, harmonization of diverse national policies would not only be difficult, but even undesirable. Perhaps the way to craft bioethics public policy or to predict the outcome of bioethics policy debates might be to start not with an ethical analysis of the technology itself but with a political analysis of the particular national or regional governmental setting. In other words, is it possible that sweeping political forces—historical, cultural, structural, legal—and the political philosophies they create may over-determine the outcome of bioethics public policy debates?

This is not to suggest that analyzing the duties of parents or physicians, or investigating the difference between withholding and withdrawing care, or studying the questions of distributive justice that underlie access to health care, is pointless; it is merely to suggest that such analyses are inadequate to justify what we ought to do or to explain why we will pick certain bioethics policies over others. Furthermore, ethical analysis, while relevant, does not represent a superior or more pure approach to public policy. Politics is not merely an irritating constraint that prevents us from doing what is right; it is, rather, an embodiment of other values best discussed in terms of political philosophy rather than moral philosophy, and these values, too, must guide our decisions.

Unique histories yield unique phenomena. The anti-biotechnology and anti-euthanasia forces, though present in all countries, are particularly robust and influential in Germany because of its troubled history of Nazism. While eugenics was most certainly a phenomenon in the United

States, it has far less hold on American politics because it was not part
of a national trauma (although it is frequently cited by anti-choice activists because of its association with early birth-control movements). One
of the few areas in which the issue of eugenics has really had traction in
the American national debate is in its intersection with racial equality, a
phenomenon that itself reflects the unique history of the United States,
in which slavery is our original sin.

Other factors are more economic or cultural than historical. In countries with national health care, for example, doctors are part of a system
of service prioritization. As a result, they are part of the management
team, and, along with economists, philosophers, and administrators, they
will form an alliance that decides which services are owed to patients and
which are not (Charo 1995). In the United States, by contrast, physicians
often work in fee-for-service settings, and they become allies with their
patients in a consumerist demand for more patient autonomy in purchas-
ing services. This is a phenomenon with obvious implications for fields
of medicine that straddle the "disease"/"lifestyle" distinction, such as
infertility treatment and cosmetic reconstruction. And, concomitantly, by
characterizing themselves as purveyors of medical services rather than as
arms of an overarching health-care system, physicians in the United States
also seem to feel freer to make claims of personal autonomy and conscience by declining to sell their services to single women seeking infertility treatments, married women seeking contraception, or rape victims
seeking the "morning after" pill (Charo 2005a, 2007).

Another American phenomenon is the tortured state of its women's
movement, torn between a view of women as the perennial victims of
men, the state, and the medical profession—a view based on a documented history of overly aggressive advances and sometimes punitive
practices in the areas of reproductive medicine—and a view of women
as rational, autonomous beings, deserving of choice even when they,
along with men, are challenged by the race- and wealth-based inequities
of our educational, employment, and health-care systems. This division
helps to explain the often surprising alliances, such as that between
long-standing critics of IVF and the anti-choice opponents of cloning
research. It also explains the difficulty the women's movement has had
in developing a coherent set of consensus positions on individual choice

in the context of genetic testing, sex selection, or gene therapy, with fear of victimization warring with fear of overreaching moralistic governmental regulation.

There are structural forces in play as well. These include not only the difference between national and federal systems (which accounts in part for the variations in embryo research policies among states in Australia and the United States, as contrasted with the more monolithic policies of the United Kingdom, Korea, and France), but also the difference between representative and parliamentary systems. Parliamentary systems insulate individuals from the vagaries of intensely local popular votes by embedding them in party lists set by party leaders and in an electorate that votes predominantly by party rather than by individual personality. The reactions to the Terri Schiavo case might have been much different if individual members of the House of Representatives did not fear the ten-second spot ("And he voted to kill poor Terri Schiavo") in the next election cycle. And this need for each politician to play to a particular segment of the electorate rather than to his party's leadership is magnified in settings that require primaries. One reason the debate on funding of stem cell research was so potent in the 2004 elections was that, having captured a prominent share of public awareness, it highlighted the degree to which the Bush administration had chosen to play to its extremes (Brownstein 2005).

Legally, the presence of a bill of rights sets American jurisprudence and bioethics public policy debates apart from those in other countries. Even where there are good reasons for governmental restrictions on a technology, the Bill of Rights and its penumbra give individuals a claim that, in certain areas, restrictions must meet a far more stringent test of justification. The result is a legal system that specifically favors the individual over the collective, the dissenter over the majority, and the eccentric over the conformist, at least with respect to free speech, reproductive choice, and other fundamental rights. In a sense, this is an approach that favors intergenerational concerns, such as the long-term viability of peaceful regime change, over intragenerational concerns, such as the most efficient or popular legal ordering for this time.

And in the 2003 Supreme Court case of *Lawrence v. Texas* (539 U.S. 558, 2003), in which a Texas anti-sodomy statute was challenged as unconstitutional, Justice Anthony Kennedy went even further, declaring that the fact that "a State's governing majority has traditionally viewed a particular practice as immoral" is "not a sufficient reason for upholding a law prohibiting the practice," and therefore that "the Texas statute furthers no legitimate state interest which can justify its intrusion into the individual's personal and private life." This was a profound statement in favor of moral libertarianism, denying the very existence of a legitimate role for the government in directing purely personal, intimate behaviors on the basis of prevalent moral codes. Without question, if this position were widely adopted in other cases and political debates, it would profoundly differentiate the United States from more communitarian forms of government.

At a still deeper level, our solutions to dilemmas in bioethics are answered by reference to political philosophy as well as politics. The pre- and post-Enlightenment debate over moral relativism is at the heart of Justice Kennedy's cry for governmental restraint in morals regulation. And the Enlightenment highlighted other fundamental differences in political approach: logic versus faith in reasoning through dilemmas; optimism versus pessimism regarding the improvability (though not perfectibility) of the human condition; embracing versus resisting the serendipitous changes wrought by economics, technology, and science. The conservative bioethics movement knows this, and has for a number of years been working quite deliberatively and effectively to create interlocking think tanks, journals, and commissions that adopt a consistent political philosophy to undergird a series of specific bioethics analyses of topics, including assisted suicide, medical futility, female sexuality, embryo research, and abortion. In March 2005, for example, the *Washington Post* broke the news that then-Chair Leon Kass and others appointed by George W. Bush to his "President's Council on Bioethics" (PCB), purportedly acting as private citizens and not as members and staff of the council, had been meeting privately for months to develop and circulate a memorandum with a conservative bioethics agenda for Bush's second term (Weiss 2005)—an agenda broader and more ambitious than ongoing congressional efforts because, as the memorandum read: "We have today an administration and a Congress as friendly to human life and human dignity as we are likely to have for many years to come. It would be tragic if we failed to take advantage of this rare opportunity to enact significant bans on some of the most egregious biotechnical practices." (source: http://blog.bioethics.net) But of course, this was not really news. Many of the specifics outlined in the memorandum had already appeared in *The New Atlantis*, a journal of conservative thought on technology and society. And William Saletan, a regular on www.slate.com, had been reporting on the Bush bioethics council's efforts to build political coalitions to undermine support for research on embryonic stem cells (Saletan 2005).

As an intellectual movement, bioethics constantly transforms itself. Long under the purview of theologians, it was snatched away by secular philosophers in the 1970s, swallowed up by physicians and lawyers in the 1980s, and infused with humanists in the 1990s. Now it is being balkanized into its liberal and conservative wings, and has come into its own as a major player in public policy debates. No longer restricted to academic conferences and advisory boards, it has even developed organs of aspiring power, such as the "bioethics defense fund," which offers white papers, policy briefings, and legal briefs on "any human life issue."

It would be absurd to imagine that bioethics discussions aimed at influencing public policy can ever be free of politics. Such "public bioethics" discussions take place on major governmental and non-governmental advisory boards, and the appointment of members to such committees is certainly the focus of political interest from time to time (*Chronicle of Higher Education* 2003; Marshall 1994; Irving undated). But the Bush bioethics council has been dogged by allegations of bias, litmus tests, and undue involvement with partisan politics (Siegal 2005; Caplan 2004; Cook 2004; Smith 2004; Holden 2004; Stolberg 2002; Bottum 2002). And very importantly, though often overlooked in public commentary, the staffs of presidential commissions were not selected with political agendas in mind.

Aggressively political agendas in public bioethics commissions are associated with failure, whether measured in terms of public policy influence or simply in terms of public credibility. As one congressional study put it: "Successful commissions were relatively free of political interference, had flexibility in addressing issues, were open in their process and dissemination of findings, and were comprised of a diverse group of individuals who were generally free of ideology and had wide ranging expertise." (U.S. Congress 1993)

The overtly conservative agenda outlined by the former executive director of President Bush's bioethics council, Yuval Levin, and implemented in many of the council's reports, has been pursued in a manner that is consistent with the attitudes associated with Leo Strauss, who followed a Platonic philosophy that emphasized the importance of elites as leaders of both discussion and society:

> Strauss believed that the essential truths about human society and history should be held by an elite, and [h]e held that philosophy is dangerous because it brings into question the conventions on which civil order and the morality of society depend. This risks promoting a destructive nihilism. According to Strauss, the relativism of modern American society is a moral disorder. . . . "Moral clarity" is essential. (Pfaff 2003)

Another distinguishing characteristic of this council is the degree to which its agenda includes rejection of moral relativism, recognition of evil, attention to curtailing "runaway scientism," and "unbridled technological advance," and incorporation of religious values in discussions aimed at shaping biological research and modern health care in ways that avoid "dehumanization" (President's Council on Bioethics, About the Chairman).

But while the PCB's work features much discussion on the subject of human dignity, scant attention is paid to what political philosophy teaches is the significant difference between arguing something is unethical and arguing that it is (or ought to be) prohibited by federal law. Its report on assisted reproductive technologies, for example, handles these issues in twelve paragraphs in the introductory chapter (President's Council on Bioethics 2004).

But attention to political philosophy is precisely what is needed to make bioethics analysis relevant to public policy. The debate, in large part, is not over whether a particular technology is absolutely good or absolutely bad. Most bioethicists would agree that nearly all technologies are fundamentally disruptive to society, because they make possible new choices and new interpersonal relationships. Thus, the question is not whether they promote change but whether the government can and should halt or curtail such changes.

As one critic of Leon Kass, formerly the chair of the PCB, put it:

One of Kass's primary intellectual influences is . . . Hans Jonas . . . one of the first bioethicists, who advocated a "heuristics of fear" to help stave off biomedical advance. . . . While at times he quotes Plato and Aristotle, Kass has more frequently argued on the basis of *Brave New World*. (Mooney 2001)

This strain of thinking has long characterized both Kass and the bioethics council he led, a kind of dystopianism strongly linked to a suspicion of all that tinkers with the "natural" world:

In 1962, political philosopher Leo Strauss, guru of many neoconservatives like Kass (and an influence on other bioethics commission members like Francis Fukuyama), wrote . . . that science risked upsetting the "natural order." . . . Strauss and Kass argue that we can know our place through a device Kass calls the "wisdom of repugnance," the gut instinct that tells us right from wrong. (Hall 2002)

This wisdom of repugnance has some of its roots in the work of other notable bioethicists, such as the aforementioned Hans Jonas (1982), who worried about the future of humanity in an age of scientific discovery. Kass has said that he has been deeply influenced by Jonas, and scholars describe both Jonas and Kass as skeptics leery of unforeseen ills in technological gains, especially concerning biomedicine (Kukis 2001).

The membership and the reports of the PCB demonstrate a widely shared vision about the perils of scientific advances, a wariness born of pessimistic views of human nature and a "resulting worldview [that] tends to Manichaenism—the notion that the world consists of a permanent struggle between the forces of good and evil, light and dark (an idea that also accords very well both with the thinking of the Christian right and, not to mention, of Bush himself)" (Lobe 2003).

According to Jonathan Moreno and Sam Berger (2007, 8),

Neoconservatives worry about the changes wrought by biotechnology because they doubt the potential for progress. Carrying their own scars from failed attempts to change the world, neoconservatives such as Gertrude Himmelfarb argue "progress is not always lovely," but rather "is unpredictable and undependable." Such neoconservatives believe that supporters of biotechnology are making the same mistakes as communists did 50 years ago in thinking that they can create a better society. William Kristol and Eric Cohen, former editor of the neoconservative journal *The New Atlantis*, explicitly draw the comparison to Marxists, arguing that supporters of biotechnology exhibit "altogether an odd mixture of the hubris of the medical researcher seeking to lead his fellow men beyond nature, and the sentimentality of the post-Communist romantic, who seeks in genetic science man's new hope for building a kind, just and liberated heaven on earth."

The ideas linking the neo-conservative journals, think tanks, and writers lie at the heart of both bioethics and more general discussions of political philosophy. Beyond the shared dark vision of science and fear of social change wrought by technological innovation is a shared belief in the permissibility or even imperative of governmentally enforced morals regulation. Whether it is parental choice about whether to embrace or avoid the birth of a child with genetic disease, or individual choice to hasten or delay death, or the intimate choice of homosexual couples to marry or not, a basic divide exists between those who call for the government, based on pure majoritarian will (often itself driven by majoritarian theology), to limit or ban such choices and those who take a more liberal (in the libertarian sense of the word) approach to morality, with its concomitant reticence on the part of government to interfere except to protect vulnerable third parties from concrete harms.

Yuval Levin, the former acting executive director of the PCB, is a senior editor of *The New Atlantis*. In the inaugural issue he wrote that "among the more prominent peculiarities of our politics in recent years is that something called 'bioethics' has become a key conservative priority." After describing the general angst brought on for some as they witness science uncovering the origins of life, behavior, and even consciousness, Levin (2003) notes that "the resulting intellectual and political activity has melded some of the interests of the pro-life movement with those of conservatives more concerned with the general culture and its institutions, and it has formed, through that combination, an altogether plausible conservative program" and asserts that "the present task of a conservative bioethics, therefore, must be to develop and articulate a coherent worldview—to put meat on the bones of loosely defined terms like 'human dignity' and 'Brave New World' and turn ethical disquiet into public arguments."

It is in this most fundamental of culture divides that the special characteristics of neo-conservative bioethics emerge. In the widespread attachment to a world view that is suspicious of technological advance, opposed to moral relativism and moral pluralism, determined to identify moral absolutes, and open to an increased permeation of religious values into public policy and bioethics analysis, neo-conservative bioethics (and the Bush PCB) appear to reflexively endorse the view that science is a threat to both society and government. Indeed, the former PCB chair himself wrote:

Science essentially endangers society by endangering the supremacy of its ruling beliefs. . . . Science—however much it contributes to health, wealth and safety—is neither in spirit nor in manner friendly to the concerns of governance or the moral and civic education of human beings and citizens. Science fosters and encourages novelty; political society, governed by the rule of law, cannot do without stability. Science rejects all authority save the truth, and prefers skepticism to truth . . . ; the political community requires trust in, submission to, and even reverence for its ruling beliefs and practices. (Kass 1985)

Perhaps this is merely the first and best articulation of a cultural divide in the bioethics world that has been brewing for years. This divide, between those who celebrate the transformative power of science and those who fear it, is both broad and profound. It is broad because it reaches into many other areas of national debate. It is hardly a leap to move from asserting that each child has a human right to be conceived by both a man and a woman (as the PCB does in its recent call for a legislative prohibition on any form of conception other than fertilization by sperm and egg) to asserting that each child has a human right to a father and a mother, a statement with obvious implications for debates on the structure of the family and access to state-sanctioned marriage contracts. And it is profound because it is a divide that reflects competing fears, with one group most fearful of the social change wrought by technology and the other most fearful of the oppressive overreaching of a government bent on controlling those changes.

More fundamental, the conservative and neo-conservative bioethics is the latest attack on the Enlightenment, the movement grounded in seventeenth-century Europe and that was the very basis for the American experiment of the eighteenth century. Enlightenment thinkers believed that human reason could be used to combat ignorance, superstition, and tyranny and to build a better world. Their principal targets were the entanglement of state and religion and the domination of society by a hereditary aristocracy or other elites. The PCB developed by the G. W. Bush administration and its interlocking journals, conferences, and funders represent an effort to reintroduce religion as the basis for public policy, pessimism, and fear as the bases of technology assessment, and elitist discussion and morals regulation as the bases of governance. It is a movement perhaps best dubbed the "endarkenment."

The Enlightenment tolerance for moral relativism is a call for government restraint in the regulation of morals. The Enlightenment also highlighted logic versus faith in reasoning through dilemmas, optimism versus pessimism regarding the improvability (though not perfectibility) of the human condition, and embracing versus resisting the serendipitous changes wrought by economics, technology, and science (Charo 2004). And it is to the Enlightenment, and its inverse, the endarkenment, that one might turn in a fresh discussion of a topic that has received a recent small resurgence of attention, one that lies at the center of some of these debates—the freedom to do research.

Consider the following words, written by Galileo:

And who can doubt that it will lead to the worst disorders when minds created free by God are compelled to submit slavishly to an outside will? When we are told to deny our senses and subject them to the whim of others? When people devoid of whatsoever competence are made judges over experts and are granted authority to treat them as they please? These are the novelties which are apt to bring about the ruin of commonwealths and the subversion of the state. (Newman 1956)

On June 29, 2006, the Vatican announced that it would excommunicate any woman, scientist, or politician who participated in or facilitated research on embryonic stem cells (Moore and Highfield 2006). Galileo would not be surprised. After all, he was threatened with excommunication merely for endorsing the Copernican view that the Earth, and by extension humanity, is not the center of the universe.

But at least the Vatican's basis for excommunication this time was its assertion that embryos are the moral equals of women, patients, and babies, and as such may not be killed to benefit others. Similar purportedly secular calls to ban certain areas of research and to jail the scientists who pursue them are based on little more than fear of social disruption and a debasement of our "humanity," a fear far more akin to the Vatican's earlier efforts to stifle research that challenged fundamental cultural, political, and theological beliefs.

Although still tentative, recently there has been increased attention paid to the question of whether the First Amendment, which is commonly understood to protect speech against government censorship, ought to protect basic science research. Beginning in 1978 with a seminal article by John Robertson, and more recently with additional articles focused on cloning research in particular (Macdonald 2005; Francione 1987; Spece and Weinzierl 1998; Delgado and Millen 1978; Favre and McKinnon 1981; Ferguson 1979), the question has been raised: Can scientific research be viewed as a form of protected "speech"?

The First Amendment does not protect only words spoken aloud or written on paper. It also protects "expressive conduct," such as the wearing of a piece of clothing that is intended to convey a message. Indeed, expressive conduct is so well protected that it precludes government bans on the burning of the very flag that represents a country whose constitution protects expressive conduct.

In analyzing First Amendment protection of research as free speech, legal scholars have usually asked whether it ought to be protected because the action (research) is a precursor to the speech (publication of results), but such arguments have failed to gain traction. Others have focused on the publication of scientific information, but not on the underlying research. A different approach would be to ask whether the research is in itself a form of expressive conduct, i.e. conduct that sends a message.

Research on the origins of the species, such as that conducted by Darwin, could thus be viewed as implicitly conveying a message, that message being a rejection of extant religious and cultural views about the singularity of humans and a substitution of a world view in which humans are part of a continuum of the animal world. Indeed, according to Goethe, Copernicus's work was perceived as having just such an effect:

Of all discoveries and opinions, none may have exerted a greater effect on the human spirit than the doctrine of Copernicus. The world had scarcely become known as round and complete in itself when it was asked to waive the tremendous privilege of being the center of the universe. Never, perhaps, was a greater demand made on mankind—for by this admission so many things vanished in mist and smoke! What became of our Eden, our world of innocence, piety and poetry; the testimony of the senses; the conviction of a poetic—religious faith? No wonder his contemporaries did not wish to let all this go and offered every possible resistance to a doctrine which in its converts authorized and demanded a freedom of view and greatness of thought so far unknown, indeed not even dreamed of. (Landry, undated)

In the modern world, research on the origins of life, such as synthesizing artificial chromosomes, using cloning to mimic the process of natural fertilization, or combining human and non-human cells to form hybrids and chimeras, challenges both what Wesley Smith calls "human exceptionalism" (which Smith contends is the bedrock of human rights) and what former head of the Bush bioethics council calls "the meaning of being human" (Smith, Secondhand Smoke, http://www.wesleyjsmith .com/blog/; Kass 2002). More recently, human "enhancement" research and neuroelectronics has been singled out as a threat to our species and a step toward "transhumanism." Any scientist who pursues these lines of research is, implicitly, rejecting the religious and political view that humanness is and must remain beyond dispute, beyond change, and beyond exploration. Of course, to meet the test for constitutional protection, such a scientist must not only implicitly make a statement of protest against extant cultural or religious views, but must actually intend to convey that protest in a form comprehensible to others.

Experimentation is not, by nature, necessarily a form of expressive conduct, as defined in 1974 by the Supreme Court in *Spence v. Washington* (418 U.S. 405, 1974). In *Spence*, which concerned a "peace symbol" taped onto a flag, the Supreme Court identified two criteria for determining whether conduct would be treated as a form of protected expression: Was there an actual intent to convey a message? Was it likely that the message would be understood? Strictly interpreted, this would preclude treating experimentation as a form of expressive conduct, short of a scientist actually announcing that he or she undertook the work in order to disprove a religious doctrine or to undermine a majoritarian view of morality.

But if the spirit (even if not the current jurisprudence) of the First Amendment supports the notion that challenges to prevailing wisdom are precisely the forms of expression that deserve protection from government restriction, ought not these emerging areas of research be protected? And if challenging the very meaning of being human, and the position of humanity within the universe, is not one of the purest forms of political expression, then what is? While most scientific experiments would rightly be considered non-expressive, some, such as those described above, certainly are, whether due to the intent of the scientists performing them or due to the effect they have on those who observe them. And if a subset of scientific experiments is regarded as protected expressive conduct, then prohibitions must be supported by more than mere fear of social disruption, and regulations must be narrowly tailored to accomplish legitimate state purposes, such as protection against concrete harms to the environment or identifiable persons. Mere speculation about harm in the future is insufficient. So, too, is the argument of "dual use," i.e. the argument that the technology can be used for both good and evil. Dual-use dilemmas exist for many technologies, including nuclear fission and recombinant DNA techniques. The answer to the dual-use dilemma lies in narrowly tailored regulation, not in prohibition or excessively broad regulation.

Now both First Amendment jurisprudence and bioethicists of all political stripes agree that "means" can be regulated regardless of the ends to be pursued. For speech, this means reasonable time and manner restrictions to prevent harm to third parties. For research, it is the basis for regulations to guard against everything from viral outbreaks to the abuse of research subjects. In the debate over non-reproductive "research cloning," however, the immediate subject of "harm" is an embryo, something explicitly not recognized as a legal person under current Supreme Court jurisprudence or historical common law. Thus, its protection cannot suffice to justify government bans on basic research, if that research deserves First Amendment protection. And concerns over dual use (e.g. use for human reproduction that threatens the physical health of children) justify narrowly tailored regulation but not prohibition of the technique. Similarly, calls to prohibit hybrids by President Bush in his January 2006 State of the Union address and to prohibit chimera research by Senator Brownback in his 2005 legislative proposals would seem to need more than fear to justify anything more than public-safety and animal-welfare regulation.

Bioethics discussions often conflate arguments about why a technology might lead to untoward consequences with calls for outright prohibition, giving inadequate attention to the equally important arguments about the moral and possibly the legal right to pursue all avenues of research provided that no concrete harm is done along the way. Those from both the political Left and the political Right calling for increased attention to social control of science would be well advised to consider the words of Thomas Jefferson: "Liberty is the great parent of science and of virtue; and a nation will be great in both in proportion as it is free."

Works Cited

Anderson, Ryan T. 2007. The end of the stem-cell wars: A victory for science, for the pro-life movement, and for President Bush. *Weekly Standard*, December 3.

Bottum, Joseph. 2002. Opinion journalism at the *Post*. *Weekly Standard*, January 18.

Brownstein, Ronald. 2005. On filibuster and stem cells, GOP bears pain of compromise. *Los Angeles Times*, May 30.

Caplan, Arthur L. 2004. Council lacks balance. *OB GYN News* 11, no. 39: 9.

Charo, R. Alta. 1995. Le penible valse hesitation: Fetal tissue research review, and the use of bioethics commissions in France and the United States. In *Society's Choices*, ed. R. Bulger et al. National Academy Press.

Charo, R. Alta. 2000. Principe de precaution, bioethique, et role des conseils publics d'ethique. In *Les Cahiers du Comité Consultatif National d'Ethique pour les Sciences de la Vie et de la Santé*. Paris: Comité Consultatif National d'Ethique pour les Sciences de la Vie et de la Santé.

Charo, R. Alta. 2004. Passing on the right: Conservative bioethics is closer than it appears. *Journal of Law, Medicine and Ethics* 32, no. 2: 307–320.

Charo, R. Alta. 2005a. Realbioethik. *Hastings Center Report* 35, no. 4: 6–7.

Charo, R. Alta. 2005b. The celestial fire of conscience. *New England Journal of Medicine* 352, no. 24: 2741–2743.

Charo, R. Alta. 2007. Health care provider refusals to treat, prescribe, refer or inform: Professionalism and conscience. American Constitution Society white paper series. http://www.acslaw.org.

Chronicle of Higher Education. 2003. Bush appoints new Advisory Committee on Human Subjects. January 17.

Cook, Michael. 2004. Embryo-centrism and other sins: The unceasing, unfair complaints of the Kass council's critics. *Weekly Standard*, September 20.

Delgado, Richard, and David R. Millen. 1978. God, Galileo and government: Toward constitutional protection for scientific inquiry. *Washington Law Review* 53: 349–371.

Eckenwiler, Lisa, and Felicia F. Cohn, eds. 2007. *The Ethics of Bioethics*. Johns Hopkins University Press.

Favre, David, and Matthew McKinnon. 1981. The new Prometheus: Will scientific inquiry be bound by the chains of governmental regulation? *Duquesne Law Review* 19: 651–703.

Ferguson, James R. 1979. Scientific inquiry and the First Amendment. *Cornell Law Review* 64: 639–665.

Francione, Gary L. 1987. Experimentation and the marketplace theory of the First Amendment. *University of Pennsylvania Law Review* 136: 417–512.

Hall, Stephen S. 2002. President's Bioethics Council delivers. *Science* 297, no. 5580: 322–324.

Holden, Constance. 2004. Researchers blast U.S. Bioethics Panel shuffle. *Science* 303, no. 5663: 1447.

Hunt, Alan. 1999. *Governing Morals: A Social History of Moral Regulation*. Cambridge University Press.

Irving, D. What is wrong with this picture? http://www.all.org.

Jonas, Hans. 1982. *The Phenomenon of Life: Toward a Philosophical Biology*. University of Chicago Press.

Kass, Leon. 1979a. Ethical issues in human in vitro fertilization, embryo culture and research, and embryo transfer. In Vitro Fertilization, Appendix, Ethics Advisory Board, U.S. Department of Health, Education and Welfare, May 4.

Kass, Leon. 1979b. "Making babies" revisited. *The Public Interest* 54: 32–60.

Kass, Leon. 1979c. A conversation with Dr. Leon Kass: The ethical dimensions of in vitro fertilization. AEI Studies 242, American Enterprise Institute for Public Policy Research.

Kass, Leon. 1985. *Toward a More Natural Science: Biology and Human Affairs*. Free Press.

Kass, Leon. 1997. The end of courtship. *The Public Interest* 126: 39–63.

Kass, Leon. 2002. *Life, Liberty, and the Defense of Dignity*. Encounter Books.

Kass, Leon. 2008. Defending life and dignity: How, finally, to ban human cloning. *Weekly Standard*, February 25.

Krauthammer, Charles. 2007. Bush took high road on stem cells. *Chicago Tribune*, December 3.

Kukis, Mark. 2001. White House bioethicist a cautious skeptic. United Press International, August 20.

Landry, Peter. Nicholas Copernicus (1473–1543). http://www.blupete.com.

Levin, Yuval. 2003. The paradox of conservative bioethics. *The New Atlantis* 1: 53–65.

Lobe, Jim. 2003. What is a neo-conservative anyway? Inter Press News Agency. http://www.ipsnews.net.

Lowry, William E., et al. 2008. Generation of human induced pluripotent stem cells from dermal fibroblasts. *Proceedings of the National Academy of Sciences* 105, no. 8: 2883–2888.

Macdonald, Barry. 2005. Government regulation or other "abridgements" of scientific research: The proper scope of judicial review under the First Amendment. *Emory Law Journal* 54: 979–1091.

Marshall, Eliot. 1994. Rules on embryo research due out. *Science* 265, no. 5175: 1024.

Mooney, Chris. 2001. Irrationalist in chief: The real problem with Leon Kass. *American Prospect*, September 23.

Moore, Malcolm, and Roger Highfield. 2006. Vatican vows to expel stem cell scientists from Church. *Telegraph* (U.K.), June 30.

Moreno, Jonathan, and Sam Berger. 2007. Biotechnology and the New Right: Neoconservatism's red menace. *American Journal of Bioethics* 7, no. 10: 7–13.

Nakagawa, Masato, et al. 2008. Generation of induced pluripotent stem cells without Myc from mouse and human fibroblasts. *Nature Biotechnology* 26, no. 1: 101–106.

Newman, James Roy. 1956. *The World of Mathematics*. Simon & Schuster.

Park, In-Hyun, et al. 2008. Reprogramming of human somatic cells to pluripotency with defined factors. *Nature* 451, no. 7175: 141–146.

Pfaff, William. 2003. The long reach of Leo Strauss. *International Herald Tribune*, May 15.

President's Council on Bioethics. About the Chairman. http://www.bioethics.gov.

President's Council on Bioethics. 2003. Being human: Readings from the President's Council on Bioethics. http://www.bioethics.gov.

President's Council on Bioethics. 2004. Reproduction and responsibility: The regulation of new biotechnologies. http://www.bioethics.gov.

Robertson, John A. 1978. The scientist's right to research: A constitutional analysis. *Southern California Law Review* 51: 1203–1301.

Saletan, William. 2005. Oy vitae. March 11. http://slate.msn.com.

Siegal, Nina. 2005. Bioethics, Bush style. *The Progressive* 5, no. 69: 24.

Smith, Wesley. 2004. Staying human. *National Review*, June 14.

Spece, Roy G., Jr., and Jennifer Weinzierl. 1998. First Amendment protection of experimentation: A critical review and tentative synthesis/reconstruction of the literature. *Southern California Interdisciplinary Law Journal* 8: 185–228.

Stolberg, Sheryl Gay. 2002. Bush's advisers on ethics discuss human cloning. *New York Times*, January 22.

Takahashi, Kazutoshi, and Shinya Yamanaka. 2006. Induction of pluripotent stem cells from mouse embryonic and adult fibroblast cultures by defined factors. *Cell* 126, no. 4: 663–676.

U.S. Congress, Office of Technology Assessment. 1993. Biomedical ethics in U.S. public policy. http://www.wws.princeton.edu.

Weiss, Rick. 2005. Conservatives draft a "bioethics agenda" for president. *Washington Post*, March 8.

Yu, Junying, et al. 2007. Induced pluripotent stem cell lines derived from human somatic cells. *Science* 318, no. 5858: 1917–1920.

4

Bioethics: The New Conservative Crusade

Kathryn Hinsch

The explosive field of biotechnology has raised more challenging social and moral questions than we can currently answer. Whole new worlds of issues are arising, with profound implications for how we think of ourselves and each other as human beings. The complexity of the science combined with real-life scenarios that were beyond our collective imagination decades ago leave us struggling—our society is woefully unequipped to understand biotechnology's effects, process the implications these advances may have for daily life, and then make laws to control usage, safety, and efficacy.

People fundamentally fear change, suffering, and death, and these three things are at the heart of most bioethical issues. Court battles over custody of frozen embryos, the desirability of "designer babies," and extreme life extension have us reeling, yet the institutions on which we have traditionally relied for guidance on difficult moral questions—organized religion, government, and the academy—have failed to keep pace with the science or societal implications underlying the issues. When it comes to bioethics there are few acknowledged leaders or spokespeople to whom the public can turn.

What's more, the media are having difficulty sorting out how to cover bioethical issues. Should health editors cover them? Do they belong in the technology section? Should religion columnists write about them, or business and political reporters? On top of that, journalists face the difficulty of covering complex moral issues in a soundbite-driven world.

The fact that bioethics has indivisible ties to public policy adds another layer of complexity. Many new technologies are still in their infancy, so there is great opportunity for shaping policy and, perhaps more important, influencing public opinion. Make no mistake, this is a race, and whoever succeeds in shaping public opinion first will profoundly affect society for decades to come.

However, at this rate the race might be easy to call. Conservative organizations that currently influence public opinion and public policy have been strategically staking out territory in a range of bioethical issues from embryonic stem cell research to end-of-life care for 10–20 years. They are positioned perfectly to frame the discussions, define the vocabulary, and, ultimately, forge public policy on bioethical issues.

Driving the Bioethical Agenda

Conservatives currently understand what is at stake better than progressives do; they see that driving bioethical debate is critical to building a society based on their values and their worldview. With little tolerance or respect for competing belief systems, conservatives see bioethics as a way to extend their anti-reproductive-freedom, anti-science, and narrowly defined religious political agenda. They use bioethics to galvanize their base, gear up the troops for battle, divide progressives, and polish their image as protectors of society's values.

While some bioethical questions currently debated may be seen as premature, speculative, and far-fetched, the positioning of issues by conservatives is shaping the way that people think about bioethics. Even if specific issues do not in fact come to pass in the way they are currently being imagined (such as "designer babies"), there will almost certainly be analogous developments. Conservatives may not be describing literally what will happen, but they are creating the frame, language, and conventions for how people will think about these issues as they arise.

Conversely, progressive activity in the area of bioethics is poorly funded and narrowly focused, and it lacks a unified philosophical framework. Although some argue that progressives dominate academic bioethics, most scholars are, understandably, strongly disinclined to work from an explicit ideological perspective. The progressive organizations that have added bioethics to their agenda are the reproductive-rights groups, which are ill-equipped to carry a broader "progressive bioethics agenda." Traditionally, groups that champion reproductive rights, such as Planned

Parenthood, have been viciously attacked by the right and have had to spend valuable resources defending against the erosion of reproduction freedom (such as the Supreme Court's 2007 ruling in *Gonzales v. Carhart*). Under such circumstances it is difficult for them to respond to emerging bioethical issues while continuing to carry out their primary mission.

Conservative Bioethics: Philosophical Framework

At the core of bioethics is the ultimate power struggle for the control of life and death. One of the best synopses of the conservative perspective on bioethical issues was captured by R. Alta Charo in her observations on the President's Council on Bioethics: "[I]n its widespread attachment to a neo-conservative world view that is suspicious of technological advance, opposed to moral relativism and moral pluralism, determined to identify moral absolutes, and open to an increased permeation of religious values into public policy and bioethics analysis, this council and its leadership appear to reflexively endorse the view that science is a threat to both society and government. . . ." (2004, 307)

An explicit conservative bioethics framework is emerging. In the January–February 2006 issue of the *Hastings Center Report*, Eric Cohen, a prominent conservative bioethicist, details that framework in his article "Conservative Bioethics and the Search for Wisdom." He reports that this framework is based on the conservative vision of human dignity and the moral ideal of equality (Cohen 2006, 44–56). Many progressives believe that if the conservative vision Cohen espouses is realized, it will effectively eliminate reproductive freedom, shut down free scientific inquiry, and intrude upon families' ability to make their own end-of-life decisions.

Bioethics: A Conservative Priority

Conservatives have been neither shy nor coy about their intentions to explicitly drive a bioethics agenda. These are not just a few scholars debating obscure philosophical ideals; they are aggressively moving to action.

In *The New Atlantis*, a conservative journal on technology and society, Yuval Levin, a fellow at the conservative Ethics and Public Policy Center and former chief of staff of the President's Council on Bioethics, noted how bioethics has become a conservative priority in "The Paradox of Conservative Bioethics." This article, published in 2003, makes their stance quite clear:

Among the more prominent peculiarities of our politics in recent years is that something called "bioethics" has become a key conservative priority. . . . Some American conservatives have long shared the concerns that animate bioethics. The pro-life movement has always worried deeply about the treatment of the unborn by scientists and doctors, and many conservatives have, through the years, been interested in various issues surrounding medical ethics, illicit drug-use, assisted suicide, and other social and cultural matters that have much to do with modern science. But it was not until fairly recently that bioethics emerged as a general and prominent category of concern for the American right. . . . This trend, together with several sensational recent advances in biotechnology, has sent bioethics toward the top of the agenda of the American right. President Bush's first prime-time address to the nation was about his new policy on the funding of embryonic stem cell research. Human cloning has been prominent on the congressional agenda for much of the past two years. And a substantial portion of the intellectual energy of the conservative movement has been devoted to the cause of a new bioethics. (Levin 2003, 53)

The conservative effort to drive a bioethics agenda became even more explicit in a leaked memo outlining a strategy for President George W. Bush's second term. The memo, written by Leon Kass in February of 2005 and posted on http://blog.bioethics.net on March 9, 2005, included these aggressive assertions:

The purpose of this memo is to outline a bold and plausible "offensive" bioethics agenda for the second term. What we hope to achieve: new biotechnologies challenging human freedom, equality, and dignity are arriving at an accelerating pace, especially in the domains of assisted reproduction and genetic manipulation. And yet there are currently no boundaries or protections in federal law to help us confront the challenges they pose. . . . We have today an administration and Congress as friendly to human life and human dignity as we are likely to see for many years to come. . . .

Rick Weiss broke the story of the bioethics agenda in the *Washington Post* on March 8, 2005:

Frustrated by Congress's failure to ban human cloning or place even modest limits on human embryo research, a group of influential conservatives have drafted a broad "bioethics agenda" for President Bush's second term and have begun the delicate task of building a political coalition to support it.

The loose-knit group of about a dozen people—largely spearheaded by Leon R. Kass, chairman of the President's Council on Bioethics, and Eric Cohen, editor of *The New Atlantis*, a conservative journal of technology and society—have been meeting since December. Their goal, according to a document circulating among members and others, is to devise "a bold and plausible 'offensive' bioethics agenda" to replace a congressional strategy that has been "too narrowly focused and insufficiently ambitious. . . .

While writing an explicit memo on bioethics strategy for a presidential administration may be new, conservatives weighing in on bioethics issues with presidents is not. Both the Center for Bioethics and Human Dignity (a fundamentalist Christian bioethics center) and the United States Conference of Catholic Bishops have been closely following and submitting reports on cloning and human stem cell research to presidential commissions as far back as Clinton's National Bioethics Advisory Commission in the late 1990s (Eiseman 2003, 37, 91).

The Players in Conservative Bioethics

To understand the current state of bioethics public policy, I surveyed 150 top conservative and progressive organizations, including think tanks, religious and philanthropic advocacy groups, legal and political groups, and university-based organizations, to evaluate the extent of their activity on bioethics issues. What I found is that conservatives have broad outreach channels to support their philosophical goals. The following is a top-line assessment of some of the most influential conservative organizations based on the following set of criteria: they include bioethical issues as part of their agenda, they have a significant constituency base, they promote an identified political agenda or set of values, they use a multi-issue approach to bioethics, and they are based in the United States.

Traditional Conservative Bioethics Centers

Let's begin with the groups that have bioethics as their primary agenda. The top conservative bioethics centers include the National Catholic Bioethics Center (NCBC), the Center for Bioethics and Culture (CBC), the Bioethics Defense Fund (BDF), and the Center for Bioethics and Human Dignity (CBHD). While their tactics and audiences may differ, they share a unified philosophy based on conservative religious values, and they are active on a range of bioethical issues. Though some may have modest resources, it is wrong to conclude that they are insignificant players in the world of public policy. These well-established bioethics centers have strong advocacy outreach programs that are interlocking and supportive of each other and, most important, are filling a void left by progressives. Here is a closer look at who they are and how they accomplish their goals.

The National Catholic Bioethics Center   The NCBC is the leading voice on conservative bioethics at the state and national level. Established in 1972 and headquartered in Philadelphia, the NCBC is funded by the Roman Catholic Church, individual contributions, and private foundations. It has a broad constituency and affiliations with over 80 U.S. dioceses, as well as numerous foreign dioceses, and it works closely with the United States Conference of Catholic Bishops. The NCBC is run by a staff of 14, including both clergy and laypeople, but because it is an official program of the Church the NCBC is not required to publish its financial data.

The NCBC's website notes that it is "engaged in education, research, consultation, and publishing to promote and safeguard the dignity of the human person in health care and the life sciences" and that "its message derives from the official teaching of the Catholic Church: drawing on the unique Catholic moral tradition that acknowledges the unity of faith and reason and builds on the solid foundation of natural law."

The NCBC focuses on stem cell research, end-of-life issues, genetics, reproductive technologies, cloning, organ transplantation, assisted suicide, and euthanasia. Its activities include a monthly column called "Making Sense out of Bioethics" that appears in various diocesan newspapers across the country as well as on the NCBC's website. A telling example is "Sperm for Sale," an article on sperm and egg donation posted on the NCBC's website on March 1, 2006. Written by Tadeusz Pacholczyk, a priest who has a doctorate in neuroscience from Yale, the column informs readers that Catholic teaching believes that "donating to sperm or egg banks violates something fundamental at the core of our own humanity" and that it "dissociates us from the deeper meaning of our own bodies and gravely damages the inner order of marriage." In addition, the NCBC publishes numerous books and two journals, *Ethics & Medics* and *The National Catholic Bioethics Quarterly*. The NCBC also holds several education conferences and bioethics consultations.

The Center for Bioethics and Culture Although the smallest player in the world of fundamentalist Christian bioethics, the Center for Bioethics and Culture is gaining prominence as an opinion maker in the area of reproductive technology and stem cell research as a result of innovative use of technology (blogs, online polls, e-newsletters) and unconventional coalition building on issues such as the perils of egg donation.

Established in 2000 and headquartered in Tiburon, California, the CBC had revenues of $130,000 in 2006. According to the philanthropic research website www.GuideStar.org, the CBC's mission is very clear: it "exists to engage, inform, challenge and mobilize a broad array of cultural influencers and shapers, in the interest of preserving human dignity, especially among the most vulnerable; and of fostering the development of a human-affirming conscience to be defended in an increasingly biotechnical society and culture."

While the CBC has changed its mission statement since 2005 to exclude specific references to religion and religious values, its pro-life political stance is still very much at the forefront of its work, only now it has cleverly repositioned itself as "pro-human." For example, it promoted a June 2006 event called "Biotech Century Series: Taking Life, Making Life, and Faking Life" with the following battle cry: "In the biotech century we can expect to face human cloning, stem cell research, reproductive technologies, genetic engineering, and end of life issues. No longer is pro-life centered on the abortion debate. In the twenty-first century, we need to be pro-human!"

The CBC's "Manifesto on Biotechnology and Human Dignity" seeks a ban on all human cloning (reproductive as well as therapeutic) and on inheritable genetic modification. This manifesto, published on www .cbc-network.org, is signed and supported by 32 prominent conservative leaders, including Chuck Colson (Prison Ministries), James Dobson (Focus on the Family), Robert H. Bork (American Enterprise Institute), and Paige Comstock Cunningham (Americans United for Life). Other activities of the CBC include lectures, seminars, conferences, and advising churches on bioethics issues.

The Bioethics Defense Fund   The Bioethics Defense Fund, formerly known as the Catholic Worldview Institute, is one of the newest players on the scene. The BDF, established in 2005, is headquartered in Scottsdale, Arizona. Its 2005 revenues were $350,000. Issues on its agenda include cloning, abortion, women's health, euthanasia, and assisted suicide. The BDF is the force behind the innovative "Cures Not Clones" project (detailed at www.bdfund.org), which that sought to tell "the truth behind the federal 'Taxpayer Loan to Clone' scam that seeks to use your tax dollars to fund unethical and ineffective destructive human embryo research."

According to www.GuideStar.org, in addition to the "Cures Not Clones" project, the BDF's activities in 2005 included "PowerPoint seminars developed by the BDF legal team and presented at 11 universities, four legislative policy briefings, three church-sponsored events, and four civic organization events. BDF lawyers were quoted in 50 print articles, interviewed on 11 radio programs, appeared as bioethics experts on three television programs and had four opinion pieces accepted for publication in a major national journal."

The Center for Bioethics and Human Dignity   With its extensive position papers, podcasts, articles, books, and bibliographies, it isn't hard to discern where the CBHD is coming from on any given issue. Established in 1994 and headquartered in Bannockburn, Illinois, the CBHD posted 2006 revenues of $790,000.

The CBDC's website states this mission: "Recognizing that biblical values have exercised a profound influence on Western Culture, the Center explores the potential contribution of such values as part of its work. In 1993, more than a dozen leading Christian bioethicists gathered to assess the noticeable lack of explicit Christian engagement in the crucial bioethics arena. This group sponsored a major conference in May

1994, 'The Christian Stake in Bioethics,' and concurrently launched the Center for Bioethics and Human Dignity."

The CBHD works on such issues as managed care, end-of-life treatment, genetic intervention, cloning, stem cell research, reproductive technologies, euthanasia and suicide, and bioethics and the Church. Its activities include development of educational materials, conference sponsorship, and a speaker's bureau. While those activities may seem unexceptional, CBHD Board Member C. Christopher Hook made it clear at a conference on "Bioethics Nexus: The Future of Health Care, Science, and Humanity" that for the CBHD bioethics engagement is war. In his opening statement at the plenary session, titled "The Future of Bioethics Engagement," Hook said:

We need to be thinking of bioethics engagement in a Marshall sense because we are engaged in a war—not a new war—it is a war that the church has been fighting since the Lord convened it 2000 years ago. It is a war against deceit, ignorance, hatred, sin, degradation, commoditization, and all the many things the enemy concocts to demean, destroy, and deprecate human dignity. So the bioethics engagement is the bioethics war for our stewardship of God's creation and our promotion of true and genuine human flourishing." (Hook 2007)

Pro-Life Groups Extend Missions Beyond Abortion

In addition to conservative bioethics centers, pro-life organizations have extended their missions to include bioethics. These groups bring the emotional intensity of the abortion debate and, most important, they bring the vast resources of the pro-life movement.

Americans United for Life   My research led me to one of the most active of the pro-life organizations devoting resources to affecting bioethics public policy at a state level: Americans United for Life (AUL). According to an article by Jeanne Cummings in the *Wall Street Journal* on November 30, 2005, AUL was founded in Washington in 1971 by two doctors who were worried that the growing women's rights movement was giving strength to a push to legalize abortion. AUL employs five constitutional lawyers and an administrative staff, and its 2004 income was nearly $1.4 million. Now headquartered in Chicago, it has 501(c)(3) status as a non-profit devoted to "civil liberties" advocacy.

In 2004, AUL began to make a concerted effort to transform from a single-issue-focused, pro-life, anti-abortion organization to one that encompasses a broad array of bioethical issues. This strategy is reflected in a revised organizational template, in press statements, in programs, in issues addressed, and in a legislative agenda. Cummings reported in the same *Wall Street Journal* article that AUL describes itself as a "non-profit, public-interest bioethics law firm." As far as I can tell, it is the only organization of its kind, progressive or conservative.

AUL positions itself as "the brains" of the pro-life movement. It uses model legislation, *pro bono* legal services, and public education as its primary tactics. Many in the reproductive-rights movement are aware that AUL is a well-established, well-funded, pro-life law firm dedicated to overturning *Roe v. Wade*, but what may not be as well known is its recent work in shaping public policy on bioethical issues that include euthanasia, physician-assisted suicide, embryo research, genetic engineering, and human cloning.

While AUL's website looks like dozens of other pro-life conservative websites, further digging revealed its startlingly sophisticated use of model legislation on bioethical issues targeted at the state legislative level. The purpose of this model legislation is to provide everything a state legislator would need to quickly and easily draft, defend, publicize, and push through legislation to promote a conservative bioethics agenda. With this "legislation in a box," AUL is primed to have a major influence on the direction of bioethics policy and the framing of the public debate.

The Center for Bio-Ethical Reform

One of the better-financed pro-life groups operating under the guise of "bioethics" is the Center for Bio-Ethical Reform, which in 2005 had revenues of $2.7 million. Established in 1991, the CBR is headquartered in Lake Forest, California and has satellite offices in Ohio, Missouri, Oregon, South Dakota, and Tennessee.

On its website, the CBR ties its activities directly to "establish[ing] prenatal justice and the right to life for the unborn, the disabled, the infirm, the aged and all vulnerable peoples through education and the development of cutting edge educational resources."

The CBR's "cutting edge" educational materials include a graphic video with a close up of a woman's vagina depicting an abortion in progress and bloody photographs of 11-week fetuses. While it has appropriated the term "bioethical," so far its activities focus primarily on stopping abortion.

The American Life League   The American Life League (ALL) states on its website that it established the American Bioethics Advisory Commission (ABAC) in 1998 to "defend the human being, his innate dignity, and his unique nature" and "not allow the rush toward bioethical tyranny to proceed unchecked."

The ALL (headquartered in Stafford, Virginia), with 2005 revenues of $7.5 million, claims to be the largest grass-roots pro-life educational organization in the United States. It positions itself as different from other pro-life, pro-family groups because it is "opposed to all abortion, contraception and other threats to the human person and the family. This total protection approach separates us from many of the other major groups. ALL will not support abortion-related legislation that contains exceptions for rape, incest, life of the mother, fetal deformity or other such condition."

The ALL's issues include stem cell research, cloning, reproductive technologies, euthanasia, genetics, eugenics, and personhood. Its activities include "Rock for Life" (a youth outreach program), a "Crusade" against pro-abortion Catholic members of Congress, and STOPP (a program dedicated to defeating Planned Parenthood). It also produces bioethics position papers and news alerts.

Traditional Think Tanks and Bioethics

While it is interesting that there are numerous centers and pro-life groups pushing a conservative bioethics agenda, what is even more compelling (and possibly more dangerous) is that well-established conservative think tanks that traditionally focused on broad economic, social, and foreign policy issues have added "bioethics" to their political agendas.

Conservatives are using their existing infrastructure of think tanks to drive awareness, energize their constituencies, and support a unified bioethics agenda. They see bioethics as a way to extend their conservative agenda, and they have an eager audience; when people are confused and afraid, they will seek the more conservative option. Their constituents are looking to them to provide education and counsel on these issues.

These conservative think tanks conduct research and analysis like universities, without the distracting students (or the need for academic integrity). They focus on issues that are politically relevant, while not burdened by the need to create "original thought" as a professor would. They also have strong marketing arms, so their programs and policy recommendations can reach a broad audience, relying heavily on repetition of messaging to push policy to adoption.

An especially active Washington-based think tank that has added bioethics to its agenda is the Ethics and Public Policy Center (EPPC). Founded in 1980, the EPPC had revenues of $2.4 million in 2005. According to www.GuideStar.org, its overall mission is to promote debate on domestic and foreign policy. In 1998 it was the first organization to sponsor a conference on neuroscience and ethics, now considered one of the "hottest" fields in bioethics. The EPPC has created a program called Biotechnology & American Democracy, which has put a conservative spin on issues such as embryo research, abortion, and transhumanism. In 2003 it launched a journal, *The New Atlantis*, as "an effort to clarify the nation's moral and political understanding of all areas of technology." It also publishes an annual "Bioethics Agenda" that contains position papers on issues from preserving boundaries between human and animals to nanotechnology.

The EPPC is just one of many conservative think tanks with impressive cash reserves that are now focused on bioethics. Among the other Washington-based organizations are these:

• The Heritage Foundation, founded in 1973 (2004 revenues: $74 million). It considers itself "the number one Washington think tank in researching and marketing conservative ideas to the President, Congress, the media and state and local governments."

• The American Enterprise Institute for Public Policy Research, founded in 1943 (2005 revenues: $35 million). It refers to itself as "one of America's largest and most respected think tanks."

• The Federalist Society for Law and Public Policy Studies, founded in 1983 (2006 revenues: $8 million). It strives to "promote intellectual diversity in the legal profession and throughout the legal community."

• The Family Research Council, founded in 1983 (2006 revenues: $10.9 million). It seeks to "reaffirm and promote nationally, and particularly in Washington, D.C., the traditional family unit and the Judeo-Christian value system upon which it is built."

• The Discovery Institute, founded in 1996 (2005 revenues: $3 million). It promotes "thoughtful analysis and effective action on local, regional, national, and international issues."

The approaches these think tanks take to bioethics differ because of their target audiences. For instance, The Discovery Institute underwrites promotional book tours for conservative bioethicists and writes op-ed pieces. The American Enterprise Institute recently held a lecture for its membership on "How to Think about Bioethics and the Constitution." The Heritage Foundation held a conference on "A Consumer's Guide to a Brave New World" with Discovery Institute bioethicist Wesley Smith. The Federalist Society has been active at various United Nations sponsored activities and worked to influence UNESCO's Declaration on Universal Norms on Bioethics. The Family Research Council has formed a Center for Life and Bioethics whose mission is "to inform and shape the public debate and to influence public policy . . . in law, science, and society," and has an extensive set of policy papers on its website.

Because of their political connections, massive funding, creative outreach programs, and large constituent bases, conservative think tanks have been extremely effective in shaping public opinion and policy and are destined to play an important role in bioethics public policy.

Bioethics and Religious Ministry

Conservative religious ministries are using their vast resources to drive a conservative bioethics agenda. One interesting example is Prison Fellowship Ministries (PFM), founded in 1976 (2005 revenues: $54.6 million). PFM's mission as stated on its website had originally been to "seek the transformation of prisoners and their reconciliation to God, family, and community through the power and truth of Jesus Christ, and the transformation of believers as they apply biblical thinking to all of life, enabling them to transform their communities through the grace and truth of Jesus Christ." Then, in 1991, PFM decided to expand its mission to include an advocacy arm called The Wilberforce Forum, whose goal is "to help Christians approach life with a biblical worldview so that they can in turn shape culture from a biblical perspective. Using the talents of leading Christian thinkers and writers, we seek to help Christians think and live Christianly not only in church and family circles, but also in the public square."

One initiative of The Wilberforce Forum is the Council for Biotechnology Policy, whose mission is to develop biblical perspectives on new, often difficult issues (e.g., nanotechnology and artificial intelligence) and to emphasize the significance of human dignity as the cornerstone of other important social and cultural values. According to the Forum's website, the primary goal is to be "a resource for those who want to understand bioethics from a Christian perspective. It's a non-partisan, nonprofit coalition of academics, ethicists, and scientists that addresses issues such as cloning, stem cell research, and the use of human embryos."

Another ministry that has added bioethics to its mission is the highly influential Focus on the Family (FoF). As stated on its 2004 tax filing, FoF is a "nondenominational religious organization whose primary objective is to spread the Gospel of Jesus Christ by helping to preserve traditional values and the institution of the family." Based in Colorado Springs, and with 2004 revenues of $140.3 million, its goal, according to www.GuideStar.org, is to "reach 400 million people in 170 countries through radio; produce and distribute resources in 36 languages and 81 countries; conduct 196 'impact projects' (such as abstinence training) in 75 countries and have associate offices in 35 nations."

Focus on the Family posts a series of mini-white papers for its vast audience on "How would God have us respond" to bioethical issues, positioned in the following way on the FoF website:

"Bioethics" is a term coined nearly 30 years ago to describe the process of examining the ethics involved with emerging medical technologies and biological research. Bioethics is interdisciplinary, bringing together medicine, law, philosophy and public policy. Bioethics covers a wide range of topics, including abortion, reproductive technologies, genetic intervention, stem cell research, physician-assisted suicide, and end-of-life medical decisions.

Practically speaking, bioethics often touches us where we hurt the most: in illness, infertility, and unexpected medical (and sometimes moral) difficulties. For Christians, bioethics is also an attempt to know what the Bible says about these issues and how God would have us respond.

But Focus on the Family's activities are not limited to bioethics position papers. In 2006, it mailed a brochure titled "Women's Voices Against Cloning: Exploiting Women in the Name of Science" to 90,000 voters in Missouri to stop "The Stem Cell Research and Cures Initiative" that would ensure Missourians the right to pursue stem cell research. Contrary to what the FoF brochure stated, the initiative explicitly banned reproductive cloning. Although that particular FoF effort was unsuccessful—the initiative was approved by Missouri voters on November 7, 2006—FoF is well positioned and well financed enough to be a major player in the field of bioethics public policy.

The Vatican (or the Holy See) is arguably the most powerful of all religious ministries when it comes to influencing public opinion. The Holy See recently announced that it was preparing a document on new bioethical questions posed by advancing biotechnology, updating the 1987 instruction *Donum Vitae*, signed by then-Cardinal Joseph Ratzinger (now Pope Benedict), that dealt with the ethical side of such topics as artificial reproduction and embryonic research. As reported on Zenit Online on January 30, 2007, Archbishop Angelo Amato described the new document as a *Donum Vitae II* and said that it is "not conceived to abolish the preceding one, but to address the different bioethical and biotechnological questions posed today, which at that time were still unthinkable." This document could profoundly shape public opinion on bioethics, as the Catholic Church's worldwide recorded membership now exceeds a billion, or about 17 percent of the population.

Exploiting Film, Television, and Literature

While traditionally the humanities have been seen as the purview of progressive thought, it is the conservatives, not the progressives, who are capitalizing on the increased interest in bioethics in cultural media, with conservative bioethics centers leading the way.

Bioethical issues and themes are showing up with more frequency in film, television, and literature. Not since the early days of space exploration have we seen the general public's interest in science-related issues so piqued. *Million Dollar Baby* dealt with euthanasia and the right to die, *The Constant Gardener* with informed consent and drug trials in developing countries, *The Island* with human clones for replacement parts, *Gattaca* with genetic engineering, *The Sixth Day* with cloning, *Minority Report* with neuroengineering, and *Bicentennial Man* with artificial intelligence. Popular television programs such as *Law and Order*, *House*, and *Grey's Anatomy* have tackled informed consent and genetic testing. In literature, Margaret Atwood showed us the perils of bio-engineering in *Oryx & Crake*, Jodi Picoult took on savior siblings in *My Sister's Keeper*, Kazuo Ishiguro focused on cloning in *Never Let Me Go*, and Michael Crichton dealt with genetics in *Next*.

Clearly filmmakers, television producers, and authors are not setting out to drive a conservative bioethics agenda; they are setting out to create compelling story lines and adequate narrative tension. Most of the recent works in this vein focus on the perils rather than the promise of bio-technology—and it is the perils that conservatives use to promote their agenda. These aren't just talking points; conservative bioethics centers have made popular movies and books major vehicles for their message.

The Center for Bioethics and Human Dignity and the Center for Bio-ethics and Culture both have comprehensive and ongoing movie and book review sections on their websites, with titles such as "Bioethics at the Movies: Review of *Minority Report*" and "The Banality of Evil: A Review of Kazuo Ishiguro's *Never Let Me Go*."

The following excerpt from the Center for Bioethics and Human Dignity's review "What Not To Do: A Review of the Film *Million Dollar Baby*," by Matthew Eppinette, is a prime example of using a creative work to make an ideological point:

Million Dollar Baby is very direct in its depiction of struggle, suffering, and isola-tion; it will leave you unsettled, perhaps shocked. In a culture that pushes for greater acceptance of euthanasia, suicide, and assisted suicide, perhaps this is a subject about which we need to be unsettled. Every human life needs proper support and care, and faith communities uniquely are equipped by the Holy Spirit to respond. *Million Dollar Baby* is a superb vehicle, portraying a "what not to do" message.

Not content to merely leverage popular culture, both centers have also developed extensive libraries of podcasts and video content. One recent video produced by the Center for Bioethics and Culture, titled *Trading on the Female Body*, uses the controversy over egg donation to suggest that embryonic stem cell research will harm women.

Leveraging the power of popular culture is a compelling strategy and one that engages the public in a visceral and dramatic way. As O. Carter Snead, associate professor of law at the University of Notre Dame and former general counsel for the President's Council on Bioethics, noted at a conference of conservative bioethicists and medical professionals, "culture and art are powerful tools," and "a movie such as *Gattaca* can drive home the dehumanizing aspects of technology more so than a very sophisticated philosophical lecture could" (Snead 2007).

Funding the Conservative Bioethics Movement

How is all this conservative bioethics activity being financed? In October 2003, the CBC posted on its website an article titled "Who Will Be the Next Joe Coors of Bioethics?" Joe Coors, of the Coors Brewing Company, was the businessman who provided the initial seed money of $250,000 (equivalent to $1.16 million today) for one of the first conservative think tanks, the very influential Heritage Foundation. The article was a plea for conservatives to step up funding of bioethical political initiatives, and the message apparently got through; according to the Foundation Center Online, in 2004 and 2005 a number of foundations, including the R. Smith Foundation, the William K. Warren Foundation, Grace Foundation, the Arthur DeMoss Foundation, and the Ochylski Family Foundation, all made contributions of between $25,000 to $195,000 to conservative bioethics centers.

Conservative foundations are also quietly and strategically funding high-profile cases with a bioethics agenda in mind. An example is the case of Terri Schiavo, which involved a family dispute regarding the removal of artificial nutrition and hydration of a woman in a persistent vegetative state. Jon B. Eisenberg, who served as one of the lead attorneys for Terri's husband, Michael Schiavo, describes how conservative foundations funded the legal and public-relations wars of the Terri Schiavo case:

> It is a story of money—vast sums of it. Hundreds of millions of dollars are funding a conspiracy—there is not better word for it—to take away our right of personal autonomy. In the Schiavo case, the money trail leads from a consortium of foundations with $2 billion in assets, through organizations created by leaders on the religious right such as Pat Robertson and James Dobson, to the lawyers and activists who fought so relentlessly to keep Terri Schiavo's feeding tubes attached. . . . The Schiavo case was just one battle in the religious Right's cultural war—a battle I watched first-hand, increasingly amazed and appalled as I followed the money trail and saw how far the culture warriors would go. (Eisenberg 2005, x–xi)

Though Terri Schiavo died on March 31, 2005, the battle is not over. To keep their cause alive, conservatives have worked to introduce legislation called the Starvation and Dehydration of Persons with Disabilities Prevention Act in South Carolina, Minnesota, Florida, Georgia, and other states.

The Power of a Unified Effort

What all these centers, pro-life groups, think tanks, and religious organizations have in common is that they are not focused on single issues; they cover an amazing breadth, including end of life, euthanasia, physician-assisted suicide, abortion, stem cell research, reproductive technologies, and genetics. They are adept at tying these together in a unified conservative framework based on a concept of "human dignity" from which they derive their position on any given bioethical issue (opposing abortion, *in vitro* fertilization, embryonic stem cell research, therapeutic and reproductive cloning, assisted suicide, and the removal of hydration/nutrition of patients in permanent vegetative states, to name a few).

Conservatives are business savvy and understand how to leverage their resources and investment for maximum impact. They also coordinate their activities to add value to them, serving on each other's boards, speaking at conferences, and developing joint promotional materials. For example, the Ethics and Public Policy Center has provided staffers to the President's Council on Bioethics; consultants to the President's Council on Bioethics also write for *The New Atlantis*; the staff bioethicist for Focus on the Family also serves as a fellow for the Center for Bioethics and Human Dignity; and the Family Research Council developed and paid for a very sophisticated (and, some speculate, focus-group-tested) pamphlet and presentation opposing embryonic stem cell research.

Conservative Political Advantages

In addition to heavy funding, a sophisticated infrastructure, and a unified philosophy, conservatives have a number of political advantages they can exploit to drive their bioethics agenda. For example, conservatives have the opportunity to undermine progressive coalitions by portraying them as divided, confused, and serving special interests, which is not entirely untrue; many of the emerging issues bring progressive values into conflict (autonomy vs. social justice, for example) and divide potential progressive coalitions. We saw this play out in the media during the recent California Proposition 71 stem cell initiative when progressive environmental and women's groups were battling with the scientific community over issues of corporate accountability, access to new technologies, women's health concerns, and the appropriateness of using state funds to support scientific research when basic needs such as education and other social services remain underfunded. In this instance, the friction between progressives gave the conservatives additional ammunition for their position against embryonic stem cell research. We should expect to see similar conflicts over technologies such as human genetic germ-line modification ("designer babies") and reproductive cloning in the next few years.

Conservatives also have the opportunity to partner on specific issues with organizations that are traditionally considered progressive. Examples of groups that conservatives may target to work with are feminist groups that oppose "social sex selection" (determining the sex of one's child through prenatal testing, and aborting fetuses or not selecting embryos that aren't of the desired sex), disability-rights groups (such as

Not Dead Yet) that are concerned about what they see as the trend toward determining that some lives are not worth of living (pre-implantation genetic selection technologies, support for euthanasia), and prominent environmentalists such as Bill McKibben (author of *Enough*) who have expressed concern about the excesses of science.

Because of their cohesive underlying theme and philosophy, conservatives have the opportunity to energize and embolden their constituencies more quickly than progressive organizations. By having a broad framework formulated from a set of agreed-upon conservative values, voters and policy makers are much more likely to know where they stand on any given issue—and understand why—without much educational outreach. By framing the issues in terms of "human dignity" (whether the issue is genetic engineering or end-of-life termination), conservatives tap into the intensity of the abortion debate. This "human dignity" framework gives also them a public-relations win; after all, who wants to stand *against* human dignity?

Progressive Opportunities

There is hope. Recent political missteps have shown signs of cracks in the conservative armor. In the Schiavo case, the majority of Americans, regardless of their position on the issue, felt it was not appropriate for the president and Congress to be intervening in a private family decision that was seen as a matter for the state courts. Realizing that most Americans favor pursuing embryonic stem cell research, members of the President's Bioethics Council have tried to advance all sorts of convoluted alternative ways of obtaining stem cells, including genetically programming embryos to die so that their cells could then be taken.

Progressives may also be able to take advantage of the blurring of traditional political alliances on bioethical issues. For example, Republican Senator Orrin Hatch, a pro-life conservative, came out in favor of embryonic stem cell research. We saw the same political crossover on a bill authorizing federal support for such research; although President Bush vetoed the bill, many prominent Republicans were outspoken proponents. In the Schiavo case, Jesse Jackson, Ralph Nader, and Rush Limbaugh all took the same side. Because of this, conservatives are beginning to fight among themselves about how best to push their agenda, and this is temporarily undermining the coalition.

In a March 8, 1997 *Washington Post* article titled "Conservatives Draft a 'Bioethics Agenda' for President," Rick Weiss wrote:

> . . . the effort to galvanize congress has already run into a major roadblock—and not from scientists, patient advocacy groups or the biotechnology industry. In an unusual instance of open divisiveness among Bush's conservative base, the nascent agenda is under attack by a variety of opponents of embryo research, including Sen. Sam Brownback (R-Kan.), who led failed efforts in the past four years to pass legislation that would ban the cloning of babies and human embryos for research.
>
> The split has some conservatives on Capitol Hill worried that the efforts of the new Congress to regulate these technologies could get bogged down in factional fighting even as several states take steps to mimic California's recent decision to expand human embryo research.
>
> "We'd like to have everybody on the same page," said David Prentice, a senior fellow for life sciences at the conservative Family Research Council. "But people have different ideas of how to do that and what the page should be."

Challenging the conservative domination of the bioethics public policy agenda won't be easy, and progressives are unlikely to match their megaphone. Progressives won't be effective by merely dismissing conservative concerns as scientifically ignorant or faith-based nonsense; what is involved is far more complicated than just saying "yes" where conservatives say "no." Progressives need to create a compelling alternative vision, a unified philosophy based on a different worldview and values. It is not too late to frame these issues and ensure that the debate is not reduced to simplistic notions.

More time must be spent thinking about what kind of world we want to live in, and then building a philosophical framework around that vision, rather than just weighing in issue by issue. We need to talk less about technology and more about values—all things moral cannot be ceded to conservatives.

Progressives must move beyond old ways of thinking about issues. Technology will change the nature of the facts and force a reexamination of our underlying belief systems, whether pro-choice, pro-environment, or pro-science. For example, when one partner in a relationship wants a frozen embryo donated to research and the other wants it implanted in another woman's womb, what is the pro-choice position? Although in favor of scientific progress, are there circumstances where it can do more harm than good?

The willingness of progressives to engage in a broad rethinking of the issues will help map out powerful and compelling positions.

The Future of Bioethics

Many new technologies—genetic testing, human genetic germ-line modification, and neural imaging, to name a few—are still in their infancy. Neither conservatives nor progressives have definitively staked out these issues, so there is great opportunity for shaping policy and determining how the issues are framed in the public mind.

If progressives continue on the present course of tackling issues one by one, they will lose the opportunity to make a difference in shaping the future of bioethics in the United States. Only once an overarching moral framework has been established, and channels for dissemination of these views have been built, will progressives be able to compete effectively with the conservative agenda.

The goal needn't be to raise the importance of bioethics to the level of social security, education, or foreign policy in the eyes of the general public. Rather, the goal should be to pursue a bioethics agenda so that progressives have a more compelling, values-based way to talk about *all* issues, not just the so-called bioethical ones.

To provide an alternative worldview, progressives must make the necessary investments in time, brainpower, and money to develop a coherent, politically savvy strategy to address a broad range of bioethical issues before this important battleground is lost, affecting us for decades, if not longer.

Acknowledgements

I would like to acknowledge the editorial contribution of Christy Raedeke and the numerous people who gave generously of their thoughts and time whose input improved the content of this chapter: Peggy Danziger, Lynn Dolnick, Jonathan D. Moreno, Lisa M. Stone, Gina Sullivan, Valerie Tarico, Susan Brown Trinidad, and Michael Yesley.

Works Cited

Charo, R. Alta. 2004. Passing on the right: Conservative bioethics is closer than it appears. *Journal of Law, Medicine & Ethics* 32: 307.

Cohen, Eric. 2006. Conservative bioethics and the search for wisdom. *Hastings Center Report* 36: 44–56.

Cummings, Jeanne. 2005. In abortion fight, little known group has guiding hand. *Wall Street Journal*, November 30.

Eiseman, Elisa. 2003. *The National Bioethics Advisory Commission: Contributing to Public Policy.* RAND Corporation.

Eisenberg, Jon B. 2005. *Using Terri: The Religious Right's Conspiracy to Take Away Our Rights.* Harper San Francisco.

Hook, Christopher. 2007. The future of bioethics engagement. Paper presented at conference titled Bioethics Nexus: The Future of Health Care, Science, and Humanity, Deerfield, Illinois.

Levin, Yuval. 2003. The paradox of conservative bioethics. *The New Atlantis* 1: 53.

McGee, Glenn. 2005. The Kass agenda: "Bioethics for the second term." http://blog.bioethics.net

Pacholczyk, Tadeusz. Sperm for sale. http://www.ncbcenter.org.

Snead, Carter O. 2007. The future of bioethics and the law. Paper presented at conference titled Bioethics Nexus: The Future of Health Care, Science, and Humanity, Deerfield, Illinois.

Weiss, Rick. 2005. Conservatives draft a "bioethics agenda" for president. *Washington Post*, March 8.

Zenit Online Editors. 2007. Vatican preparing bioethics document. January 30. http://www.zenit.org.

5

Justice That You Must Pursue: A Progressive American Bioethics

Laurie Zoloth

Ethics is the first responsibility. To think is no longer to contemplate, but to commit oneself, to be engulfed by that which one thinks, to be involved, this is the dramatic event of Being in the world . . . we are thus responsible beyond our intentions . . . committed to duty before we are free.
—Emmanuel Levinas, *The Temptation of Temptation*

Justice is the one thing you should always find
You got to saddle up your boys
You got to draw a hard line
—Willie Nelson and Toby Keith, *Beer For My Horses*

Learn to do good; Seek justice, Reprove the ruthless, Defend the orphan, Plead for the widow.
—Isaiah 17 (American King James Edition, 1995)

How are we to work for the good? What a surprising question for an academic article about bioethics, and yet how odd that it should be so surprising. For what we claim we do in the field of bioethics is to know the truth and drive the search for the good act by our questions, and by our keen discernment. Yet, for much of our new field's career in the academic medical school, we have acted as if we needed to keep out of the fray "our politics," our faith commitments, and most assuredly our passion. We thought ourselves, however, a "we" who had convictions, and whose convictions offered a default, vaguely left-wingish frame for policy.

During the Bush administration, the field of bioethics learned a lesson about praxis. The moral act is a public act, and thus an act within a polity and within a system of power that is always in contention. The power to discern the true and the good is worth a good argument. In the Bush administration, the argument deepened and became one of the central ways that culture was defined and one of the central issues of political debate. How we judged science, medical care, bodies, and theories became who we were, shifting the ethical to the ontological. Bioethics became definitional. Our colleagues who spoke from a conservative position understood this first and most clearly, and crafted a multi-volume, multi-year conservative bioethics position. There is a need, then, to respond, both to "reproach" what became in some cases "ruthless" and to "seek justice, defend orphans and speak for the widow."

Reproach

The last days of August 2001 were a dreamy, sweet time in American public life. After all, both the Democrats and the Republicans had won the presidency, and we had adjusted to that. On August 9, 2001, we watched George W. Bush create a pretty good political compromise on human embryonic stem cells that seemed to work. This pleased and distressed both sides of what had so far turned out to be the most pressing issue of our time, and of that day, unbelievable though it now seems: molecular biology. President Bush called for the favorite of all things for bioethicists, a bioethics national panel. Bush asked someone to chair it for whom we had a great deal of respect, a doctor who was a prolific and gorgeous writer about moral philosophy, Dr. Leon Kass. I was pleased with several of his choices for board members, who despite being Republicans, perhaps unlike any single person on Clinton's National Bioethics Advisory Commission I remind you, and despite the fact that some were completely unknown in the field or the history, organizations, or the texts of bioethics, were surely interesting thinkers. Since the president had declared his policy not only firmly, but based on implacable moral signs, I was not surprised when the President's Council on Bioethics (PCB) announced that it would deal not only with stem cells, a science project I knew to be in only its most tentative stages, but also with other vivid, pressing concerns in bioethics, "deep issues" we were told, "that we often ignore." How could I, long troubled by many aspects of a thin, autonomy-driven bioethics quarrels that had mesmerized the field, not be pleased by the promise of such attention? Like Leon Kass,

I was not always impressed by many aspects of modernity. I had written an obscure but Kass-like article about the falsely seductive aspects of dying one's hair or dressing like a teenager. Like any sane person, I knew that science needed careful and reflective regulation. I was hoping for good things—deep things, as it was promised on the website, like health-care reform, vaccination policies, attention to international health crises, the impact of poverty on health, or the problem of justice in the pharmaceutical industry.

It was this very affection and admiration for the idea of having a bioethics committee at the national level that said it was committed to interesting ideas that has made it so troubling, when, three years later at the annual conference of the American Society for Bioethics and Humanities (ASBH), I could only call it tragically flawed. In this chapter, I return to that original paper, for it lays out the questions that, by the last years of the Bush administration, had fully been explored—and can be assessed. It is the contention of this chapter that the conservative bioethics agenda, as expressed by that panel, did a profound disservice to the field of bioethics, to the best ideas and vision of a progressive, democratic America, and not the least, to the needs that should be paramount in bioethics: the suffering of the befallen and the desperation of the poor.

Let me be clear: It is the intellectual arc, the basic choice at the core of the PCB's work that is an error and that set the field of bioethics on a profoundly tragic path. For in nearly all of its work, the PCB's majority reports have chosen to read the text of science as metatrophic synecdoche for a modernity out of control, absent of social order or law, a world in which the sky is falling and our hands are in the air like a Greek chorus: we are certain of our catastrophe. In this world, the scientists who study molecular biology are not to be trusted, not their facts, not their dreams, not their academies. In this world, the suffering of others is not "the most important moral issue," for suffering, we were told, often is "ennobling or instructive" for us to watch. In this matter, the PCB stepped out of the reality in which I argue bioethics must remain rooted: the desperation and the need of the poor, the terrible ground of serious illness, and the existential crisis of death.

In the highly stylized world of the PCB, actual families and actual ordinary loss matter very little. What the poor know is of little value, what the rest of the world and their religions teach matters very little, and America needs to go it alone if she is morally right. In this world, core values, common sense, and gut instinct are a more reliable source of wisdom than sorting through conflicting reasons for a decision that might change with new data. In this world, Americans face a fearful future in which, unless the correct policy is followed, we will be lead fundamentally away from our liberty. In this world, the problems of the wealthiest Americans are critical to address first, for they and their power seem to be our obsession.

For the reader this will sound familiar. It is a theory for a field driven by "outcomes." It is the theory that led too much of our foreign and domestic policy; the growth of the very sort of Big Government that we most fear, one that is more akin to a powerful monarchy than a messy, Jacksonian democracy. And it is this theory—this deeply Heideggerian theoretical construction that undergirds the most ideological of the PCB's "Pentateuch" (or the Five Books that they had produced by that time—since then, they have written three more)—that pervades their book *Beyond Therapy*. It is Heideggerian on three counts: it mourns a lost pure nature; it claims that death defines us; and it is, as was said about Heidegger, "a mixture of genius and Sunday school preaching."

Beyond Therapy was a book that Dr. Kass was very proud of: it was, in fact, the only topic that he would "allow" the panel I participated in at the ASBH to talk about, and thus the rules of this talk were set. It is, alas for bioethics, neither a reasoned philosophic argument nor a coherent theological one. Rather, it is a sentimental call for a return to a romantic past, just at the moment when we needed a tough and determined bioethics that is fully prepared to cope with the future if we are to understand the complexities of biotech and participate meaningfully in the realpolitik of the future in which they must be regulated—in which our power means something.

Beyond Therapy is an ontology, a theodicy, and an eschatology. It is a Victorian, pietistic retelling of the narrative of Genesis, postulating an alternative choice: Don't touch that apple. Consider the central claim:

May our children ... continue to reap the ever tastier fruit of biotech—but without succumbing to their seductive promises of a ... better-than human future, in which we shall be as gods, ageless and blissful. ... We live near the beginning of the golden age of biotechnology. . . . Yet, not withstanding these blessings . . . we have also seen more than enough to make us anxious. . . . The powers made possible by biomedical science can be used for . . . ignoble purposes, serving ends that range from the frivolous and disquieting to the offensive and pernicious. . . . Powers . . . available as instruments of bioterrorism . . . agents of social control . . . means of trying to . . . perfect our bodies and minds and those of our children. Anticipating possible threats to our security, freedom, and even our very humanity, many people are increasingly worried about where biotechnology is taking us. (Kass 2003)

There is much to say about this long quotation, including the retelling of the Genesis story, the recasting of the researcher as the evil tempter, children as innocents, humans as fallen, and the eschatological anxiety at the heart of Kass's argument. It is this fear, a fear that drives so much of conservative bioethics, against which the contributors to this volume raise the deep hope of progressive bioethics.

First Point: The Rhetoric of Eschatology

Conservative bioethics makes much of the eschatological vision. Yet, is the complaint of conservative bioethics serious? Today looks a lot less like "a golden age" after the long panic about gene therapy and designer babies. It is not a golden age any more then many other times: eyeglasses, safe c-sections, hand-washing, and window screens were easily as revolutionary. This "end of days" narrative has a rhetorical structure as rigid as the sonnet. First part: the set up—it is a golden age! We will be as gods! But the seducer will tempt us, and we fall, gluttonous and venally, down the slippery slope out of the lost and lovely past. Final part: great power leads to great responsibility, thus, as in Superman, as in Plato, we are told that we need a wise, thoughtful elite—perhaps special doctors of some sort, say bioethicists—but beholden to absolutely no one, nor elected—and they will protect us. But is the sky really falling? Is the world about to be fundamentally changed? This is, I believe, a fundamental problem and it is a folie á deux—shared by both extreme sides in the debate, who actually believe that the future is a certainty, that we are at an eschatological moment in human history. This is an error, reflecting a rather touching sense that the future can be named, known, and spectacular. Will we actually, and in a widespread way, ever use, say, cloning, to take a popular panic, and will cloning ever really have an impact on people's reproductive practices? What about IVF (widely used, to be sure, but hardly a threat to the family or the practice of erotic sexuality)? Are we seriously debating the good or evil of immortality as if this was an actual concern?

Eschatology is a problem for progressives as well, and one cannot flee from dystopia to utopia. As a reader of the *Left Behind* novels, which I read at the same time as the PCB reports, I was struck by the parallels. Conservative bioethicists often issued proclamations as if beleaguered, as if they were members of the Tribulation Force, fighting the demonic forces of liberal thought. Much of what they feared as technology "just around the corner" has proved impossible technically (cloned humans, genetically selected children, etc.) and has largely been abandoned as research pursuits except by bioethicists, who can still be funded for researching such matters.

Second Point: The Sentiment of "Spirituality and Souls"

Much of what is discussed by conservatives in their work seems theological—guilt, sin, evil, souls—but claims religiosity without the usual work of religion, i.e., God; Law that binds you to restraint; vows of poverty or a preferential option for the poor; humility or visiting the wretched, angry sick; teaching Bible to messy, ice-cream-faced 8-year-olds; or fasting for Ramadan. "Religious views" are cited, but such citations are a stand-in for a certain form of Christianity.

To give one concrete example, for Jews across the denominational spectrum the human embryo made in a lab by cloning, or by any other means and for any intent, is just not the moral equivalent of a human person. Religious Jews do not have the same duties to it that they would have toward a human person. This is a matter of personal religious understanding guided by my actual religious interpretive community, whose Laws bind me, because this is really a matter of faith as surely as they obligate me to face the stranger and give him my bread. This will not change, and I frankly do not mind that it won't, even if most Americans believe a blastocyst to be a human person. The fact that most American believe in Christ with full conviction does not convince me that I must. This is not because I am a crass utilitarian, for I am not, nor because I am seduced by science, for I am surely not, but because I actually believe in a broken world that a good God ultimately gave to the children of man. I believe in attending to it with all my passion, and that there are Laws that guide my choices and my attention.

I want a religious voice in the very middle of the public square. In fact there need to be many such voices, and most will argue that there is a research imperative, not that we should retreat from or fear science, as, for example Fukuyama claims (2002). There is an imperative to know the world and to heal it, not for some cheap power, or an instrumental utility, or naive enthusiasm, but because one must be a *Rodef tzedek*—a pursuer of justice—and a *Rodef shalom*—a pursuer of peace. Justice. Conservative bioethicists always worry that stem cells or genetics would make us "lose our soul," but if this was at stake then why not ask the central question of the Hebrew Scripture, the New Testament, and the Koran: Can you forgive your neighbor his debt? Will you understand that his need surrounds you and that the corners of your field belong to him? Will you be present for your neighbor in his desperation?

Third Point: Life in the Aristocracy
But some have a different view of their neighbor:

The human whole, truth to tell, is never simply complete. Even at our prime and in full health, we are always under the sway of desire. We always wish to be more, to look better, to be stronger, to feel . . . happier than we are now, more accomplished and successful than our neighbors. Many of the new biotechnologies, developed originally to help those who need healing, can readily be used to help those who simply want more for themselves. (Kass 2003)

That is the human whole? What has happened to gratitude for one's breath, duty, regard for the stranger, and most of all the call of the one who is ill? This is an ethics for the guy who lives next door to a CEO. We compete with our yuppie neighbors, wanting to be more successful? Possessing of more? I think not. It is not a world like that for most—for happiness is not some game of me winning over you. For most people, surely most women, life is far more like a laundromat (perhaps not a place the PCB has experience with?). It is a matter of how to share, how to "make do" with talk over the long waits. It is a matter of raising children and feeding them when there is not enough meat in the stew. Raising children is, interestingly enough, one of the core issues of *Beyond*

Therapy. There is an odd romanticism about the meaning of childhood, along with quotes from Rousseau in which children are elegized, or claims reminiscent of 1950s television are made: "the sweetness, freshness, and spontaneity of life are available in their purest form only in the as yet unburdened young" or "modern life threatens the innocence and the simple joys of childhood" (President's Council on Bioethics 2003, 94), but it is well to remember the actual condition of France in 1755, or of black children from 1957 to 1963, when *Leave It to Beaver* was on TV. What children can the PCB be thinking of? Perhaps the same ones that live in the competitive patrician suburb, but surely not in Lesotho or South Chicago. If bioethics is to "speak for the orphan and the widow," we have to attend to far more than designer babies and fantasies of DNA-altered children wandering about the future like so many troubled teenage avatars.

I soon became interested in the vast ethical problem of foster care. If we care about the lives of children with the passion we claim in our bioethics research about genetics, then surely the poorest children living here in our country, homeless, poor, abused, and seeking justice, should attract our attention. Yet, the conservative leadership of bioethics has held not one hearing about such children or the loss of their "simple joys." One can go to dozens of websites and read the stories and see the visages of yearning children. The website called "Wednesday's Children" will send you a list of children who need homes every single week. Some are disabled, some merely poor. All are desperate. These children are disproportionately poor, black, or Hispanic. They cycle in and out of public care, typically several times, until they are 18. What are the attendant ethical issues in the system? We really do not know, for in contrast with our moral panic about genetics, bioethicists have not really studied the issue at all.

Fourth Point: There Is a Silencing of Illness
When you worry about the rich guy and your envy of him, or you fear that he will alter his children in the imagined future to be unfairly better, you are, quite literally, in the wrong room of the discourse. The room I would suggest we stay in is the clinic in Haiti or the Medicare nursing home with the staffing problem and the shortage of wheelchairs.

Bioethics ought to have a preferential option for the poor and a duty to the sick. It is not so hard, I think, to know the difference between the ill child and the vain, rich guy.

The Error of Narcissistic Bioethics

Conservative bioethics began to take on the excesses and the formulations of the new century—that the regard of the other toward me was the single most important gaze. The principle of autonomy, which was a Kantian idea based in respect for persons, has been taken to some distant Aristotelian model of dignity. What has gone awry here? Why has a basically thoughtful idea—be careful with science advances—become a Level Orange threat? It is because of the incorrect turn toward an Aristotelian principle of "dignity," which Dr. Kass tells us is related to the root for "dignitary." That is a tragic error as well, although perhaps it signals the final end of the troubled dream of autonomy itself, with all that it implies. Let me suggest an alternate principle: that of hospitality.

For Levinas, the danger of the Heideggerian impulse in large part is Heidegger's obsession with death as our defining telos. For it is hospitality, Levinas argues, not death, that actually defines and opens us away from the terrified little self that is challenged by modernity and its choice. Hospitality opens us toward the other, and not fear of the next thing that walks near our door. For Levinas, the danger of the Heideggerian impulse is fear:

The entire life of a nation . . . carries within itself . . . men who, before all loans, have debts, owe something to their neighbor, are responsible . . . and in this responsibility want peace, justice, reason. Utopia! This way of understanding the meaning of the human . . . does not begin by thinking of the care men take of the places where they want in order-to-be [but] . . . above all of the for-the-other . . . which, in the adventure of a possible holiness, the human interrupts the pure obstinacy of being. (Levinas 2000, 231)

The problem, we are reminded, is not how to be righteous. The problem, as Levinas puts it, is why is it righteous, why is it defensible to be? For Levinas, and for me, the answer must be and the justification must lie in duties that emerge from the brokenness of being itself, which pulls us not to concern for our own perfection or vague spiritual journey, but toward the needs of the other. We are pulled toward the other in her complete need not because she is a pathetic victim, but because her interruption defines and authorizes our being. Our abundant capacity is created by the way we must come up with what she needs, even in a situation of scarcity. I stand here speaking, always with more than enough, for of course I have enough to give to you in your hunger. An ethics of hospitality means an interruption of desire. It means first of all seeing the suffering presence and not prioritizing it on a list of goods you want. I am already asked when I begin to speak, and if I do not hear clearly enough, I will be interrupted by the call in the middle of the night: Where are you? Where were you to be found? Hospitality, we need to understand, is not pity, and it is not the story of a lost Eden, but of the tent of Abraham and Sarah, who live at the crossroads; who will cook for you; whose tent is lit as if by fires, by passions that are seen for miles in the desert. And in the scripture of all Abrahamic traditions, this story of hospitality is read directly before the call for justice. Yes, even justice for the most horrific abusers, for the cities of Sodom and Gomorrah; for the hospitable one will help the ones in need, even if their rescue means rescuing the sinners, as the PCB fears. Yes, we have taken the apple, and yes, we know we can insist on moral choice, even to the God of history. Hospitality leads toward justice.

Do I worry about telos? Of course, for the world to come is always at stake. That is why one must work so hard to make it come, having left Eden so long ago, and having been told with great specificity that nature must be shaped. Humans in *Beyond Therapy* are uncertain creatures, "puzzling upward pointing unit(ies)." I do not know about pointing upward, for much of life that I care about happens horizontally—making love, giving birth, bathing a child, sitting by the bedside of a dying patient. Bioethics is most free when we remember we are entirely bound to this principle of hospitality, which interrupts our opining, even our reading, and always, alas, our fine aspirations; it is the central thing that we can do.

Learning to Be Good

A progressive bioethics, we can hope, would at the very least pay as much attention to the poor as, say, the Rotary Club. Consider the following quotation:

Rotary International is the spearheading member of the Global Polio Eradication Initiative and is the largest private sector donor. It has contributed more than US $600 million to the polio eradication activities in 122 countries. In addition, tens of thousands of Rotarians have partnered with their national ministries of health, UNICEF, the World Health Organization, and with health providers at the grassroots level in thousands of communities. A polio-free world is within our grasp. (source: http://www.rotary.org)

While we in bioethics spent eight years fighting about the agenda that the PCB put forth—that science was fearfully out of control; that cloning would make us selfish, soulless, and racist; that genetically modified corn would poison us—the Rotary Club nearly ended polio worldwide. If the progressive left and bioethics had pitched in, it might have helped. What has blocked the efforts are conservative forces who told women that Western medicine would render their babies sterile and that vaccines were evil. Unlike other fields, such as medicine itself, and law with a *pro bono* requirement, bioethics has not insisted on praxis of service. We often act as if such a praxis would come at the expense of reading, or of higher philosophy.

I was a nurse for a good long time, far too long a time to read the work of the conservative bioethicists, however beautifully written and however passionately felt (for it is felt with much passion) without some sorrow, and some irony. I, like many in bioethics and in research, sat in far too many darkened rooms with parents, as their baby struggled and died, to speak of designer babies as a fearful problem. These parents would be happy in some ordinary easy way with babies that could breathe on their own. That is the hard reality of the core dilemmas in bioethics.

Alzheimer's is not a brisk fall day by a lake and the end of a long good life; it is illness, and a tragedy beyond telling. A person seeking relief from clinical depression is not looking for "mood brightening"; he is ill, and it is a loss beyond measuring. It is not a lifestyle choice to have memories so horrific that one cannot function; it is illness, and if a Stanford senior who wants to forget a wretched date sneaks a dose of a medication intended to help a person in real emotional pain, so what, really? Does it render the use of opiates by post-surgical breast cancer patients unseemly because people are addicted to heroin? Should therapy be seen as a sort of "corked bat," or should the *Beyond Therapy* uses of plastic surgery concern the burn patient? Does it cheapen us, or is it really scary, that the wealthy yuppie might use a drug intended to help a child? Tawdry, perhaps, but really scary? Does it destroy my dignity that some poor, shaken violinist will use a drug so he can have a good job and do a nice thing with music and not weep on stage? Do we really think that it is credible in a country that cannot figure out how to get flu vaccine distributed, or pay for an actual vegetable in preschooler's lunches, or housing for the poor, that MRI machines will be set up to scan the brains of unsuspecting teens in impoverished communities? Turn in *The New Republic* from Dr. Kass's letter about the threat to human dignity that "genetic engineering for designer babies" might create (Kass 1997) to the small articles about the Sudan buried in the inner pages. Designer babies? These are not the babies we should be having conferences about—these are not the baby calls in the night we should be hearing, that should wake us, now—10 minutes from now, ten decades from now, in the PCB's sci-fi future. What should waken me breathless is that my neighbor has come to my door, is begging on the streets in Philadelphia as we speak in the Big Hotel, and he needs me; that the children of the woman who is upstairs now, wiping my toilet and my floor in the Big Hotel, have inadequate health care, not the moral panic that PCB has created.

There was far more to say about these troubling ideas at the Kass Council, but ten minutes was ten minutes, and I had to end my remarks that evening at the ASBH conference. During those ten minutes, I noted in closing, 38 children had died of pneumonia, 29 of diarrhea, 20 of malaria, 15 of measles, 5 of TB, and in the next ten minutes 209 more children will die of preventable illness. Not one a designer baby, merely the ones who surround our tent of meeting, or who are in the neighborhood that surround our academic medical centers. Listen, I told my colleagues in bioethics, and listen you who are reading now. They are here, and yes, they are the ones beyond therapy that a progressive bioethics ought to be pledged to right now.

A Progressive Bioethics: The Moral Gesture in a Divided Democracy

So what of justice, and the advocacy for the orphan and the widow? There is a new chance for us, in this turning toward our American democracy. I returned to this chapter in the summer before the 2008 elections. Turning from the errors of our interlocutors in conservative bioethics, I want to end this brief chapter by calling for a new direction. Here is another text to consider:

Is not this the fast that I choose: to loose the bonds of injustice, to undo the thongs of the yoke, to let the oppressed go free, and to break every yoke? Is it not to share your bread with the hungry, and bring the homeless poor into your house; when you see the naked, to cover them; and not to hide yourself from your own kin? Then your light shall break forth like the dawn, and your healing shall spring up quickly; your vindicator shall go before you; the glory of the Lord shall be your rear guard. Then you shall call, and the Lord will answer; you shall cry for help, and he will say, Here I am. If you remove the yoke from among you, the pointing of the finger, the speaking of evil, if you offer your food to the hungry, and satisfy the needs of the afflicted, then your light shall rise in darkness and your gloom be like the noonday. (Isaiah 58: 6–10)

Much of bioethics is spent on the wrong fast, spent wailing and wearing sackcloth and ashes about a dystopic future that is not now, and may never come at all, with bioethics as a field wanting credit for all that anguish, claiming it is the "deeper" issue we should address. That is wrong. The deeper issue is not that deep—it is right in front of us. It is the poor who are naked and hunger. For a Jewish ethicist, bioethics is premised on the persistent idea that the structure of the world, its brokenness and its yearning, is the subject of our most profound attention. We place our discipline between the despair and loss of illness and the contingent possibility for redemption that can be manifest because we have good arguments about how to live as decent human beings, and because we understand ourselves as linked by duty to one another.

Progressive bioethics will be shaped not only by the core questions of Kant (What can I know? How ought I to decide? For what can I hope?), but also by three other questions I have suggested: What does it mean to be human? What does it mean to be free? What must I do about the suffering of the other?

A progressive bioethics means first of all seeing the suffering presence, the one who comes to you with all of her need. I am already asked when I begin to speak, as Levinas notes, and if I do not hear clearly enough, I will be interrupted by the call in the middle of the night—where are you?

Progressive politics is fundamentally rooted in optimism and a commitment to the power of democracy. Such a theory does not claim perfect knowledge. In the 1960s, progressives who sought to remake and redeem social organization in America based their statements on the power of Americans to have a robust public discourse, constructed not around dignity, but around solidarity, not in how I am seen by others, but how I see others before me. Consider a text from that period, the Port Huron Statement, in all of its flaws and naiveté still resonant:

In suggesting social goals and values, we are aware of entering a sphere of some disrepute. Perhaps matured by the past, we have no sure formulas, no closed theories—but that does not mean values are beyond discussion and tentative determination. A first task . . . is to convince people that the search for orienting theories and the creation of human values is complex but worthwhile. We are aware that to avoid platitudes we must analyze the concrete conditions of social order. Our own social values involve conceptions of human beings, human relationships, and social systems. We regard men as infinitely precious and possessed of unfulfilled capacities for reason, freedom, and love. In affirming these principles we are aware of countering perhaps the dominant conceptions of man in the twentieth century: that he is a thing to be manipulated, and that he is inherently incapable of directing his own affairs. . . . We see little reason why men cannot meet with increasing skill the complexities and responsibilities of their situation, if society is organized not for minority, but for majority, participation in decision-making. This kind of independence does not mean egoistic individualism. . . . Human relationships should involve fraternity and honesty. Human interdependence is contemporary fact; human brotherhood must be willed however, as a condition of future survival. (Students for a Democratic Society 1962)

Many of those who now lead bioethics came of age in the early 1960s. The need for new conversation about justice transformed our student life, and this transformation inspires us still.

"To think is no longer to contemplate, but to commit oneself, to be engulfed by that which one thinks"

This, argues Levinas, is the event of Being. It is, surely, the event of bioethics. In this chapter, I have argued f that we must not only enter the public square, but fill the square with every voice we can raise. The stakes of how we know, teach, research, experiment, exchange, succeed, and fail in science and in medicine are the stakes of American life and death. Without the voice for justice, access to care, honesty, and what we used to call "fraternity," how will we argue for health-care reform? The reason behind all the progressive agendas we can list—health-care access, and others—is simply this: we must answer the need of the suffering other, for it is the right thing to do, it is what makes us human.

Works Cited

Fukuyama, Francis. 2002. *Our Posthuman Future: Consequences of the Biotechnology Revolution.* Farrar, Straus and Giroux.

Kass, Leon R. 1997. The wisdom of repugnance. *New Republic,* June 2.

Kass, Leon R. 2003. Ageless bodies, happy souls. *The New Atlantis* 1, spring: 9–28. http://www.thenewatlantis.com.

Levinas, Emmanuel. 2000. *Entre Nous: On Thinking of the Other.* Columbia University Press.

President's Council on Bioethics. 2003. *Beyond Therapy: Biotechnology and the Pursuit of Happiness.* Government Printing Office.

Students for a Democratic Society. 1962. Port Huron Statement. http://coursesa .matrix.msu.edul.

The New Oxford Annotated Bible. 1994. Oxford University Press.

III

The Sociology of Political Bioethics

6

Professionalism and Politics: Biomedicalization and the Rise of Bioethics

Paul Root Wolpe

Bioethics is a strange field (discipline? subject? pastime?), constructed differently than most other similar academic pursuits. It is not defined by training; those who claim to be its practitioners come from a variety of disciplines in the humanities—social sciences, law, and clinical care, among others. Some come from outside academia, including clinicians, journalists, members of advocacy groups, and clergy. Many of those who write articles in areas clearly considered bioethical by those in the field do not consider themselves as "doing bioethics" or see themselves as bioethicists, while many self-described bioethicists write about things others consider outside of bioethics' domain. No central body defines or licenses bioethicists, and in that sense one becomes a bioethicist primarily by declaring oneself one. If the claimant is sufficiently convincing, or can get support from those in positions of authority, the appellation sticks, and a bioethicist is born.

One might argue that the field of bioethics (if not the people that inhabit it) is defined by subject matter. Once that might have been a useful approach. Today, however, there is much debate about what lies within its purview. Clinical care, certainly; stem cells, by general agreement; animal cloning for consumption, well, OK. But how about genetically modified endangered species? Sustainable agriculture in undernourished populations? Synthetic biology? The use in hospitals of PVC products that can leach phthalates into patients? Nanotechnology? It seems that bioethicists have been dramatically increasing their disciplinary jurisdiction, and now include anything that has a direct or indirect biotechnological connection.

The expansion of bioethics follows a similar expansion in the sophis-
tication of biotechnology that, it will be argued, is important in understanding the professionalization and increasing prestige and power of
bioethics as a social actor. From an obscure discipline, bioethics has risen
to become a central forum for debating issues of fundamental importance
to modern society. Bioethicists write columns for newspapers and other
media and appear as experts on news reports. Presidential bioethics
councils and commissions have become high-profile actors in government. Bioethicists are often invited to give public lectures or lead public
forums on bioethical issues.

Emile Durkheim was the first to argue persuasively that a society's
negotiations over what is acceptable and unacceptable, and its constantly
developing shared moral sense, define that society in contrast to others.
The values, behaviors, and range of attitudes and opinions that are
deemed acceptable in a society differentiate that society from others. In
that sense, for reasons described below, bioethics has become an important player in modern Western society's self-definition. In the United
States, no issue has symbolized that debate more than abortion; I once
heard a Pro-Life spokesperson claim that "the fight over abortion is the
fight over the soul of America." In a Durkheimian sense, she was right;
deciding that issue one way or another would be instrumental in determining which vision, which self-definition of America, will win out and
come to define the nature of the society.

In this chapter, I will argue that the professionalization of bioethics
has been a necessary part of the rise of biotechnology. Bioethics serves
the dual purpose of providing a forum for the aforementioned value
debates and legitimizing the biotechnology industry by providing a moral
justification for its products. Bioethics as a discipline therefore is torn
between its potential role as watchdog and critic and its function in
lubricating the social acceptance of biotechnologies. The position is made
even more complex by the fracturing of pre-biotechnological coalitions
(such as Pro-Life and Pro-Choice) in the face of new biotechnologies
such as stem cells. Such shifting of the political landscape makes the
positioning of bioethical positions even more problematic.

Medicalization and Biomedicalization

Sociologically, the expansion of bioethics makes sense, and corresponds with a similar expansion of biomedicine itself. For years sociologists had noticed and written about the social process of medicalization, whereby phenomena previously defined in other social realms were redefined as medical. A characteristic trait of the late nineteenth and twentieth century, medicalization turned being a drunk into being an alcoholic, being "loose" into having a sexual addiction, and shyness, distraction, and melancholy into pharmaceutical events. The process of medicalization itself continues unabated; fatness is now obesity, aging is a disease, and in some school districts up to 20 percent of kids are medicated for attention-deficit disorders.

Medicalization has many causes, but one motivation is to increase the jurisdiction, and thus the power, of medicine. The more human traits, states, and behaviors that can be redefined as medical, the more products the medical-industrial complex can churn out to address them. The expansion goes both ways. As more and more behaviors are defined as medical, more and more lay institutions are reframing themselves as quasi-medical providers. Fitness clubs, yoga studios, organic food providers, health food stores, sporting goods stories, and other commercial institutions advise customers on their health-related activities, and recast them from hobbies or leisure activities ("sports") to medical activities ("fitness").

Adele Clarke and her colleagues (2003) take the medicalization argument to a new level. They suggest that there is a new process under way that they call *biomedicalization*—a technoscientifically enmeshed form of medicalization that is fundamentally changing the nature of biological (not just medical) enterprises. Biomedicalization includes "the transformations of both the human and nonhuman made possible by such technoscientific innovations as molecular biology, biotechnologies, genomization, transplant medicine, and new medical technologies" (ibid., 162). Complex and multi-leveled, it includes the restructuring of biomedicine through computer and information technologies; the use of new clinical tools, from imaging technologies to computer-generated pharmaceuticals; the extension of medical jurisdiction to health itself, which results in individual responsibility for health and its management; privatization and globalization of research and production; highly integrated institutional structures (e.g., centralizing and standardizing patient records, insurance, etc., or open-source bioinformation); the transformation of bodies and identities through customization (e.g., cosmetic neurology, new kinds of prosthetics); and other trends.

The Rise of Bioethics in the Context of Biomedicalization

This new complex of biotechnological advances has contributed to the rise of the bioethicist. A relatively short time ago, bioethics was an obscure field, bioethics centers were scarce, and bioethical conversations happened largely within the clinical medical literature (with a smaller conversation happening in the philosophical and theological literature). Now bioethics has exploded, with centers in most major universities, degree-granting programs, a national professional organization with over 1,500 members, federal grants targeting bioethical inquiry, and Presidential Commissions in the United States and oversight bodies in many other countries dedicated to addressing, and in many cases regulating, areas under bioethical jurisdiction.

The concurrent rise of the bioethicist and biomedicalization is not a coincidence. While medicalization altered traditional categories, it did so under the rubric of medicine, a familiar and trusted social institution. Though recasting drunkenness as alcoholism was not without its social tensions (the incomplete nature of that transformation is still evident in the moral approbation clinging to alcoholism), there was little organized resistance. Substance abuse had a clear and recognized medical component, and the willingness of medicine to take a messy social problem under its wing—perhaps even find a solution—was a move that many policymakers and activists welcomed. (Interestingly, the parallel track for handling alcoholism, Alcoholics Anonymous, is still more a premedical drunkenness model than a medicalized alcoholism model, and thus AA often resists using drugs for the treatment of alcoholism even though the medical literature supports it value.) Certainly, society had to adjust to the medicalization of social categories that it had previously thought of as sinful, criminal, or willful; still, the process of medicalization unfolded over a century, and slowly embraced more and more of social life.

Biomedicalization has been swifter and much more traumatic. The rise in the rate of scientific progress itself over the last century has been claimed to be exponential rather than linear, with new technologies and human capacities to manipulate biological life emerging with astonishing speed and with much cross-fertilization (Kurzweil 2005). Absorbing and adjusting to such change is a challenge, even for a society that is used to scientific change. Biomedicalization challenges a number of embedded normative beliefs, from religious beliefs about the nature of life and death to cultural beliefs about selfhood and authenticity. The stem cell debate, cloning, and abortion challenge our conceptions of when life begins and ends; synthetic biology undermines our beliefs about what life is and who creates it; the Theresa Schiavo case and Dr. Jack Kevorkian's fame reveal our lack of consensus about when life ends and who controls that moment; Prozac and Viagra blur our boundaries between normal and enhanced functioning; and so on. The social and ethical questions that are embedded in biomedicalization are not tangential or trivial, but lie at the heart of the integration of biotechnological products into the social fabric.

At the same time, enormous social and economic resources are needed to produce new biotechnologies. Biomedical research funding alone was $94.3 billion in the United States in 2003 (Moses et al. 2005), and U.S. industry revenues in biotechnology rose from $8 billion in 1992 to $63 billion in 2005 (Lähteenmäki and Lawrence 2006). Investments in cutting-edge technologies like nanotechnology are staggering—the military alone requested $417 million for nanotechnology research in 2007 (Berger 2007).

Bioethics and Professionalization

Professionalization can be thought of as the process of internal occupational control. A number of components characterize professionalization, including the development of associations among professionals (to exchange information, provide mutual support, control practitioners, etc.); control of work (through licensure); establishing recognized educational structures to train practitioners (and to control who is recognized as a practitioner); the development of professional knowledge (through journals, websites, libraries, books, listservs, blogs, and so on); and profession-dominated worksites (e.g., bioethics centers) (Abbott 1991). Bioethics has, over the last two decades, engaged vigorously in all those activities, with the exception of licensure. While licensure (of clinical bioethicists) has been discussed and debated, there is little consensus on precisely what knowledge qualifies one to be a bioethicist, and therefore there is little consensus on what types of testing would be appropriate to grant licensure.

What is clear from the criteria listed above, however, is that bioethics has been engaged in a process of professionalization. There were only a few bioethics centers in the United States a couple of decades ago, but www.bioethics.net, a premier bioethics website, now lists more than 50, and the list is incomplete. There are more than 30 journals dedicated to various aspects of bioethics, and scores of national and international professional associations support scholars engaged in bioethical inquiry. There are more than two dozen Masters of Bioethics programs at American universities. A Google search on the term 'bioethics' garners nearly 6 million hits.

Perhaps bioethicists simply saw an opportunity, a society in need of bioethical guidance, and capitalized on it. Clearly, expertise develops when society desires it, and, as I suggested above, rapid developments in medicine and biotechnology generated anxiety that could be abated (or exploited) through the development of a profession whose job it is to define, manage, or diffuse that anxiety. Bioethics could be seen simply as a profession that grew with a public need for its services.

However, it is equally true that the biomedical and biotechnological industries need bioethics. Many of the processes and products of biotechnology challenge traditional ways of thinking about human life. For most of human history, for example, an embryo was an entity in a uterus; the object and its locus could not be decoupled. Religious belief, ethical principles, and emotional associations all evolved around a subject/object gestating in a uterus. With the advent of *in vitro* fertilization and other technologies, for the first time the living embryo could exist in a Petri dish, separate from a uterus. The question "What is that thing in the dish?" then arose (McGee and Caplan 1999). The question was not hypothetical or theoretical; millions of dollars of biotech investment depended on how society might answer that question. In such contexts, bioethicists became central figures in the desire of industry to influence the debates through bioethical expertise (Wolpe and McGee 2001).

The point here is not to claim, as some have, that bioethics was co-opted by industry. It is rather to point out that the position of bioethics was strengthened by industry, which needed bioethical discussion. The desired outcome for industry was, of course, to promote its own view of such things as the status of the embryo. But even more important, industry needed society to come to some consensus on the use of reproductive tissues so that the boundaries in which it could operate were clearly defined. Industry has a very low tolerance for regulatory ambiguity, and in the United States the development of stem-cell science was stymied by an inability to reach consensus about how embryos could be used in biotechnological research.

Professionalism and Politics

Bioethicists do not control entry into the profession the way that medicine, law, and many academic disciplines do. One can proclaim oneself a sociologist, but the claim will be extremely hard to defend if one does not have a PhD in sociology from a recognized school. Bioethicists, on the other hand, have no means to control entry into the profession. Similarly, the field has no way to expel someone from its ranks, or even to plausibly argue that a person who claims to be one is not a bioethicist. The result is a field with extremely fuzzy boundaries and little internal coherence. For example, like much of the country, bioethics can be broadly split into a liberal and a conservative camp, with a large gray area in between. While there are no clear boundaries, and on some issues people primarily in one camp or the other cross over, for a number of important issues in the ideological battles of bioethics the lines are clearly demarcated. What is also clear is that the *institutions* of professional bioethics—the national associations, the journals, and the academy, are predominantly on the liberal or progressive side of bioethics. The result is that bioethicists on the conservative side of the debate have felt the need to establish their own journals, centers, and conferences.

The ambiguity of who counts as a bioethicist has allowed critics on the right to label all bioethics and bioethicists as "liberal" and to define themselves out of bioethics. Thus one finds books (such as Smith 2000 and Stevens 2000) and articles attacking bioethics from the right by those who, by any reasonable criteria, are themselves bioethicists, but who position themselves as outsiders looking in on the enterprise. These books do not attack bioethics as a pursuit and do not suggest that moral discussion of bioethical issues is illegitimate; rather, they tend to attack establishment bioethicists as partisan, arrogant, or self-appointed and exclusionary. The identification of the term 'bioethics' with a cohort of people rather than the pursuit results in books like Jonathan Baron's *Against Bioethics*, which is, in fact, an argument not against bioethics, but for a better way to do bioethics. Similarly, critics on the right tend to be critical of bioethics, and to define everyone they disagree with as a "bioethicist" while exempting themselves and those who agree with them from that category. In his book *Life, Liberty, and the Defense of Dignity*, for example, Leon Kass writes of the "blindness" of American bioethicists, presumably exempting himself from that category even though he was, at the time, the chair of President Bush's Council on Bioethics. The result is that, despite the best efforts of the architects of bioethics' professionalization, the field remains poorly defined and split along political lines. It even penetrates organizations that are seen as bastions of bioethics, and thus, by definition (at least, by the definitions of the right), liberal. For example, the membership of the American Society of Bioethics and Humanities (ASBH)—the national professional organization of bioethicists (as well as those who study the medical humanities)—voted not to permit the ASBH to take positions on issues other than those that narrowly effect the profession (e.g., challenges to academic freedom).

In view of all the past difficulties of professionalizing bioethics, the future will see a new class of bioethicists with more firmly established professional identities. Though today the field is made up of self-proclaimed bioethicists, most of whom do not have degrees in bioethics (as no such degrees existed), the establishment of Masters programs and of

Bioethics Centers will generate trained scholars who will be able to use their credentials as validation of their status as bioethicists. Though the ambiguity of the field now prevents bioethicists from mounting a strong argument that the makeup of a commission or advisory groups does not include a bioethicist (who, after all, qualifies as a bioethicist?), the establishment of a new generation of scholars who have degrees in bioethics will make identifying "legitimate" bioethicists easier.

As a class of professionals clearly identifiable as bioethicists is created, the temptation for corporate and other interests to use those professionals for its own ends will increase. The field of bioethics may bifurcate into academic bioethicists and in-house corporate ethicists, the former being significantly more progressive than the latter. Paradoxically, as the field of bioethics increases its internal control over who can claim the label through the establishment of degree-granting programs, it may also find the profession specializing in new ways as it plays a more active role in the development of the biotechnological industry.

Conclusion

The field of bioethics is undergoing dramatic changes, and its future direction is, at best, uncertain. However, the centrality of bioethical debate to Western society's progress, the development of bioethics programs in predominantly liberal universities and bioethics centers, and the increase in the public visibility and prestige of bioethics in the public forum suggest that the influence of bioethics will grow, and that the role of progressive thought in the bioethical enterprise will have to be protected and nurtured if it is not to become the parochial posture of the academy alone.

References

Abbott, Andrew. 1991. The order of professionalization: An empirical analysis. *Work and Occupations* 18, no. 4: 355–384.

Berger, Michael. 2007. Congressional pork bloats U.S. military nanotechnology spending. *Nanowerk*, June 20 (http://www.nanowerk.com).

Clarke, Adele E., Jennifer R. Fishman, Jennifer Ruth Fosket, Laura Mamo, and Janet K. Shim. 2003. Biomedicalization: Technoscientific transformations of health, illness, and U.S. biomedicine. *American Sociological Review* 68, April: 161–194.

Kurzweil, Raymond. 2005. *The Singularity Is Near: When Humans Transcend Biology.* Viking.

Lähteenmäki, Riku, and Stacy Lawrence. 2006. Public biotechnology 2005—the numbers. *Nature Biotechnology* 24: 625–634.

McGee, Glenn, and Arthur L. Caplan. 1999. What's in the dish? *Hastings Center Report* 29, no. 2: 36–38.

Moses, Hamilton III, E. Ray Dorsey, David H. M. Matheson, and Samuel O. Their. 2005. Financial anatomy of biomedical research. *JAMA* 294: 1333–1342.

Smith, Wesley J. 2000. *Culture of Death: The Assault on Medical Ethics in America.* Encounter.

Stevens, M. L. Tina. 2000. *Bioethics in America: Origins and Cultural Politics.* Johns Hopkins University Press.

Wolpe, Paul Root, and Glenn McGee. 2001. Expert bioethics, public discourse, and the case of embryonic stem cells. In *Beyond Cloning*, ed. S. Holland et al. MIT Press.

7

The Tension between Progressive Bioethics and Religion

John H. Evans

Progressive bioethics is in formation, as an explicit alternative to a recently institutionalized "conservative" bioethics. In any initial formation process, much has to be worked out regarding the movement's aims, principles, and processes. One of the most critical issues has to do with its relationship with other institutions in American society. How would progressive bioethics relate to the state? How would progressive bioethics relate to the market? In this chapter, I will address how progressive bioethics could relate to religion—a long-standing and powerful institution in American society. This relationship cannot be invented from nothing, but rather will be heavily structured by the previous relationship between religion and bioethical debate.

History of Bioethical Debate and Religion

Jonsen (2006, 23) portrays the common wisdom among bioethics professionals as follows: "Bioethics began in religion, but religion has faded from bioethics."[1] That is the received wisdom, and it is true as far as it goes. However, we need a much more detailed understanding of this transformation to understand the contemporary challenges facing the relationship between progressive bioethics and religion.

Medical ethics has an ancient history, which we can ignore for the purposes of this chapter. For present purposes we can start in the 1960s, with the emergence of what is now called "bioethical debate." In the 1960s, myriad medical interventions and scientific discoveries were coming together to create the sense that traditional ethical boundaries were being crossed. In the 1950s the first successful kidney transplant occurred, and by the early 1960s drugs had become available to make kidney transplants more available. The first heart transplant occurred in 1967. Advances in genetics made it seem possible that old eugenic dreams of the perfection of the human species might well be possible in the near future. An early discussion of human cloning was occurring, as was the more imminent possibility of "test tube babies." People were also concerned about mind control. All these questions seemed suspiciously close to the sort of broad "meaning of life" questions that religion had traditionally had in its bailiwick. Let me be more technical in the description of these early debates. To introduce a distinction that will be important as this chapter proceeds, this early debate included a debate about what ends American society should pursue, with these new technological means at its disposal: species perfectionism? human health? individual freedom? fidelity to God? This is more generally to ask: "What is our purpose on this earth? What should be our goals?" This is the sort of debate that religion is good at. Many religious leaders at the time seem to have thought so, and many of the participants in the early debates with scientists were theologians or were explicitly religious, and many (including Joseph Fletcher, Paul Ramsey, Richard McCormick, Daniel Callahan, Al Jonsen, James Gustafson, J. David Bleich, and Fred Rosner) were among the "founding fathers" of bioethics. The new technologies raised the big questions that religion had always debated, and religion thought it had an important role to play.

Almost immediately, religious voices within the bioethical debate started to become marginalized. The religious voices that remained did not use explicitly religious language, and what made them "religious" became more difficult to discern. The common-wisdom narrative of the bioethical debate continues by saying that the reason for this marginalization was that the religious could not speak the neutral secular language that people had begun to recognize as necessary in a pluralistic society. H. Tristram Engelhardt (1986, p. 5) summarized this well:

The history of bioethics over the last two decades has been the story of the development of a secular ethic. Initially, individuals working from within particular religious traditions held the center of bioethical discussions. However, this focus was replaced by analyses that span traditions, including particular secular traditions. As a result, a special secular tradition that attempts to frame answers in terms of no particular tradition, but rather in ways open to rational individuals as such, has emerged. Bioethics is an element of a secular culture and the great-grandchild of the Enlightenment ... the existence of open peaceable discussion among divergent groups, such as atheists, Catholics, Jews, Protestants, Marxists, heterosexuals and homosexuals, about public policy issues bearing on health care, will press unavoidably for a neutral common language. Bioethics is developing as the lingua franca of a world concerned with health care, but not possessing a common ethical viewpoint.

M. Therese Lysaught (2006, 101) similarly describes the "standard narrative of the genesis of bioethics":

... its earliest origins lay among theologians, but substantive theological discourse was quickly replaced by the more advanced discourse of philosophy. ... Bioethics emerged because of the increased technologization of medicine in the 1960s and that theological ethics was pushed aside because "philosophical ethics offered the hope of resolving such questions without appealing to the faith of a particular community."

I have a slightly different version of this history. The common-wisdom narrative would explain why theologians could not use explicitly religious language and say something to the effect of, "cloning is an affront to Jesus' teaching as indicated in Matthew 5–7." However, as Jonsen says, these "early religious bioethicists ... dispensed with their outward religious appearance in order to make themselves welcome and comprehensible to the secular world" (Jonsen 2006, 34). To anticipate some crucial distinctions below, these were the types of theologians who were willing to translate their theological beliefs into a secular language for consumption in the public sphere. Given that they were speaking a secular language, why would they then be marginalized in bioethical debate? They were not obviously appealing to the faith of any one community.

My argument, expounded at length elsewhere (Evans 2002a), is that the religious participants, along with people from sociology and some other professions, were increasingly marginalized as the debate that was taken to be important moved from a debate about "which ends should we adopt" to "given a set of assumed ends, which means will maximize these assumed ends." That is, it was not the theological or transcendent discourse of theologians that made them marginal to the debate; it was what they wanted to talk about. People who appeared to be entirely secular who wanted to talk the same way, like Leon Kass, were marginalized for the same reason.

It is important to note that this was not a conspiracy. There was no meeting at which the philosophers in the debate teamed up with the lawyers to say "how do we get these theologians out of the debate." Rather, it was a gradual process that, I argue, worked primarily through cohort replacement and not through a strong change in the way any one person talked. Of course there were people who spoke more about these deeper issues in earlier years and not later years. But, primarily, the debate changed as people who wanted to speak about these deeper issues decided not to enter the public bioethical debate, and those who wanted to speak about these issues and who were already in the debate left the discussion or retired.

What happened was the creation, as Engelhardt pointed out in the passage quoted above, of what is considered to be a "universal moral language" for use in bioethics—a language not beholden to any particular religion or philosophical perspective, and thus suitable for the creation of public ethical stances in pluralistic societies. I have described this elsewhere (Evans 2000) as a series of institutionalized, taken-for-granted ends that should be pursued in medicine and science. The most common universal moral language or "common morality" is called *principlism*. It consists of four ends that we are to pursue—ends considered to be more or less universally held by citizens. These ends are autonomy, beneficence, non-maleficence, and justice. (I recognize that this is not how the designers of principlism think of principles. Rather, I am making a claim about how principles are used by average scholars.) What is critical for Engelhardt is that these four ends are expressed in an entirely secular register, and justified on secular grounds. What is critical for me is that there were only four of them, and that these ends became institutionalized as *the* ends to pursue not only in human experimentation—where they were first invented—but for all "bioethical problems." If there were still a debate about ends to pursue, there would not be only four legitimate ends to pursue. People would be debating matters like "Perhaps we should pursue this end with technology X. Perhaps we should pursue this other end with technology Y."

Once the debate was no longer about which ends to pursue, but how to maximize the four institutionalized ends, theologians lost interest. This is just not what theologians are interested in. As James Gustafson (1978, 387) has written, "it is not easy to give a clearly theological answer to a question that is formulated so that there are no theological aspects to it."

Others have described this loss of interest in debating ends similarly. Henk ten Have writes that bioethics "is dominated by an engineering model of moral reasoning and impregnated with the idea of a technical rationality, applying principles to practices" (in Lysaught 2006, 103). Indeed, during this transformation one of the founding theologians came to worry that religious ethicists were beginning to focus on these little questions. Gustafson wrote: "Should one cut the power source to a respirator for patient y whose circumstances are a, b and c? [This] is not utterly dissimilar to asking whether $8.20 an hour or $8.55 an hour ought to be paid to carpenter's helpers in Kansas City." (1978, 387)

My narrative is slightly different from the conventional account. It was not explicit theological language of the theologians that led to their marginalization, but their interest in debating ends—their interest in the "big questions" that the technologies raised. And these big questions were being displaced by a focus on the little questions in bioethical debates—little questions that flowed once the big questions (e.g., what ends to pursue) were institutionalized. I am not saying that there is no discussion about ends in these debates, or that there are no theologians in these debates, but rather there has been a tendency for "deep" questions to be marginalized.

The marginalization of religious voices from public bioethical debates was largely complete 20 years ago. Here the attentive reader will object that in debates about cloning in the newspapers, in debates over stem cells, and in debates over abortion, religious voices are everywhere. True. We therefore need to make some more distinctions. What I have been talking about is the "public bioethical debate," the debate that is held among societal elites—primarily academics who hold appointments in institutions of higher education, who serve on government commissions, and who publish in academic journals. These are the people writing the chapters in this volume.[2]

With the exception of the issue of abortion, which has involved a broader spectrum of people, these people *were* the bioethics debate on most of these "bioethical" issues until very recently. But this has changed, for two reasons. First, in recent years some religious people—explicitly distancing themselves from the mainstream institutional bioethics I have just described—have either started institutions of their own or "taken over" existing institutions, like the government ethics commission. Second, much of public bioethical debate has moved toward a more social-movement orientation, with people who are better described as activists and movement intellectuals than as academics taking a larger role.

To understand this development, and what it means for religion and progressive bioethics, we need to better understand religion in the United States. That is because, although well known yet rarely acknowledged, all of the original theologians who participated in public bioethical debate would be, by today's standards, religious liberals and not religious conservatives.[3] And, the increase in religious voices in debates about "bioethical" issues is from conservative religious voices, not liberal ones. An understanding of this difference is critical.

An Excursus on Religion in America

In the 1920s, Protestantism dominated public life in the United States. Yes, there were Jews, Mormons, and other religious minorities, but their influence on public life was even less than one would expect given their numeric representation. There were increasing numbers of Roman Catholics, but they were largely the despised immigrants of their day. They were not integrated into positions of public power; rather, the Protestant elite thought Roman Catholicism should be controlled.

In the 1920s there was a "culture war" within Protestantism between the modernists and the fundamentalists (Noll 1992, chapter 14). They clashed over many issues, but as an example, consider Biblical interpretation. A modernist could show, using modern literary methods, that if you go back to the original Hebrew the prophet Isaiah was not predicting that a "virgin" would bear a son, but that a "young woman" would bear a son. What is more familiar to most people is the willingness of the modernists to engage with the positivist science at the time, particularly Darwinism and geology, to say that the Genesis account is not literally true. Fundamentalists were, generally, Biblical literalists, fairly separatist, and socially conservative. The modernists had more social power: they were the heads of colleges, the professionals, the elected officials, the journalists, and so on. The theological descendants of the modernists are what we would now call mainline Protestants. And, to get a bit ahead of myself, these mainline Protestants are the people who in the 1960s helped found what we now call bioethical debate.

It is hard to summarize mainline Protestantism in a paragraph, but here are some generalizations we can make about it. Mainliners are not Biblical literalists. A classic mainline phrase would be "the Bible is a written account of people struggling to understand God." Mainliners are socially more liberal than other Protestants. Their denominations defended the *Roe v. Wade* abortion decision for many years, are the denominations most likely to ordain women, and are the denominations that discuss and debate gay marriages and gay clergy. They are, to put it quite simply, more liberal than other Protestants.

The fundamentalist-modernist split occurred in the 1920s. In the 1940s there was a split within fundamentalism. For those readers who are not from this conservative Protestant world this may seem like splitting hairs, but a group of people, the most famous of whom was Billy Graham, thought that the fundamentalists were too rigid and the modernists too wishy-washy. They created a new movement called "evangelicalism" as a half-way stop between fundamentalism and mainline Protestantism. We could call them liberal fundamentalists or conservative mainline Protestants. Jerry Falwell was a fundamentalist. George W. Bush is an evangelical. Again, these may seem like small distinctions if one is not from this world, but evangelicals are more liberal then fundamentalists on theological matters, like Biblical interpretation. They are also much less separatist. Both fundamentalists and evangelicals are, in my terms, religious conservatives.

A critical point for the development of religion and bioethical debate is that both the fundamentalists and the evangelicals took one lesson from their split with the modernists in the 1920s, which was that the mainline establishment was opposed to them and that they should retreat from national public life and focus on saving souls. Mainline protestant-ism continued to dominate the public sphere up until the 1960s or the 1970s. In the 1960s, lessened discrimination against Catholics and Jews meant that these groups were slowly allowed into the conversation, but only to the extent that their form of argumentation was similar to the way that mainline protestantism made its arguments. Much has been said about the protestantization of Catholicism, but let us just say that it was the more liberal Catholics who were a part of the conversation—those whose thinking most closely emulated that of mainline Protestants. But even in the 1960s and the 1970s, fundamentalists and evangelicals largely remained aloof from the public sphere, despite their fairly large numbers in terms of the overall population of the United States.

What Happened to Religion and Bioethics in Mainstream Institutional Bioethics

During the early years of bioethical debate, there were no fundamental-ists, evangelicals, or traditionalist Catholics involved in bioethical debates (as I have defined these debates) of which I am aware. Depending on how one counts, half of the population had no religious representatives in public bioethical debates. Why no fundamentalists or evangelicals? They were still in full retreat from the world. Why no traditionalist Catholics? I would argue that bioethical debate drew upon primarily academics and public intellectuals, who were primarily the liberals (like university-based Jesuits) or those who had a Vatican II orientation (like Daniel Callahan at the time).

While the bioethics debate was becoming more secular because main-line Protestants and liberal Catholics had started moving away from it, evangelicals and fundamentalists decided to re-enter public life. In the late 1970s they joined traditionalist Catholics, who were already active in the public debate over issues like abortion. This is the now familiar story of the emergence of the "religious right" in American politics (Wuthnow 1988). What is important for present purposes is that slowly, over time, the religious right did two things. First, it built up a large machinery for influencing public policy, including think tanks and social movement organizations such as the Christian Coalition, Concerned

Women for America, and Focus on the Family. Second, it became a core constituent of or took over—depending on one's view—the national Republican Party.

I would argue that public bioethical debates remained under the radar of the religious right up until the year 2000. Yes, there were skirmishes over abortion, such as the cancellation of a congressionally chartered bioethics commission due to the abortion debate (U.S. Congress 1993, 12–13), as well as some 1980s and 1990s debates about embryo research. But, by and large, public bioethical debates continued on as if the religious right did not exist, and religious conservatives more generally had not decided to become involved with public life again.

I have written elsewhere about how mainline Protestants influence public life not through the ballot box, nor through political activism, nor through grassroots organizing, but rather through elite channels (Wuthnow and Evans 2002). Mainline Protestants have influence by being the establishment, and bioethical debate, up until recently, was an establishment activity. It was some academics and a few government commissions making recommendations about what was ethical. The general public was not deeply involved. Indeed, the whole point of the public bioethical debate of the academics was to shape that debate into terms that made more sense to the state that had to regulate these activities and the scientists who had to abide by these ethical mandates.

It should be no surprise that conservative Protestants in the religious right decided that they could not influence the world through the establishment activity of the liberal Protestants, but instead would influence the world by political organizing, grassroots efforts, and elections. Not only was this an accurate assessment, it fits quite well with one of the core beliefs of evangelicals, namely that they are a beleaguered and oppressed minority (Smith 1998). Religious conservatives came to learn that the way they could influence the world was through direct political influence, because the other, more subtle avenues to power like public bioethical debate were not open to them. It was the same way these elites encouraged the organization of the religious right.

I would argue that religious liberals are more influential through elite channels than through direct political action, whereas the religious conservatives are more influential through direct political action than through elite influence. The conservatives think that this is the way to influence public life, and have developed much machinery to do so.

We then have the framework for the emergence of religious conservatives in bioethical debates. In the 1980s, the religious right more or less ignored bioethical debate because the debate was focused on issues that religious conservatives are not so concerned about. Moreover, the influence of bioethical debate on actual policy outcomes was difficult to discern. In the 1990s, bioethical issues received growing attention from religious conservatives. Then George W. Bush became president in 2001, at about the same time that a number of issues that religious conservatives really cared about also came into public view. Dolly the sheep had been born a few years earlier. Embryo research was more and more desired by scientists as it increasingly seemed that embryonic stem cells might have some therapeutic use. Bioethical debate increasingly had actual influence on what happened in the world, making it more important to fight over (Caplan 2005, 13).

Did religious conservatives then say "Hey, now that our issues are on the table, let's get our young people to get PhDs in bioethics, let's get invited to the next meeting at the Hastings Center, let's get involved with mainstream bioethics?" Of course not. Bioethical debate as an elite institution was still closed to these perspectives, because the "deep issues" that religious conservatives want to talk about were not in favor in bioethical debate. They knew that they could not win through "reasoned arguments" with bioethicists or others in the debate—or with the head of the National Institutes of Health—because these people would not accept their reasoning. So, to have influence, they started a social movement. But they already had one, at least on some of these issues, like the right-to-life movement and the machinery of the religious right more broadly.

As an example of an end run around traditional elite bioethical debate, consider how public influence was exercised on end-of-life issues. Instead of creating academic articles about the right to life of people in vegetative states and presenting these to academic conferences, the religious conservatives generated grassroots pressure to have Congress intervene in the case of a comatose woman in Florida named Terri Schiavo, ultimately passing laws to stop her ex-husband from removing her feeding tube.

Whatever your conclusion about the rightness or wrongness of their actions, it is clear that this was an enormous contribution to getting the public to see a particular ethical position about end-of-life issues.

In addition to the use of social movement organizations and political activism to influence debates about bioethical issues outside the mainstream of bioethics, the religious conservatives were also able to have a historically greater influence on a federal bioethics commission than they had ever had before. While people will debate the qualities of George W. Bush's federal bioethics commission for many years, and while it is clear that not all of the people on it were religious conservatives, or secular conservatives, certainly it was more conservative in both senses than any of its predecessors. How was this change in the character of this institution, so powerful in its agenda-setting ability, achieved? The religious right was now enough of a constituency of the Republican Party that it could demand greater adherence to its views from the Executive Branch. It is clear that Bush wanted to make the Executive Branch more amenable to the religious conservative "base" of the Republican Party, so the federal bioethics commission was more conservative. I would argue that Bush was the first modern evangelical president,[4] and he seems to have the same views on social issues as the evangelical leadership. I should note that the deliberations of Bush's President's Council on Bioethics were not explicitly religious. Rather, they were focused, more than its predecessors and particularly under the tenure of Chair Leon Kass, on secular discussions of the "deep questions"—the ends we should pursue—that religiously oriented people in these debates want to talk about.

The end result is that we now have a "culture war" in bioethical debate between the conservatives (who include evangelicals and traditionalist Catholics) and the liberals (who are not obviously religious). I should digress to say that this moment in bioethical debate was probably inevitable with the increasing democratization of the public sphere. From the 1950s to the present the story of the American public sphere is a decline in the over-influence of mainline Protestants and the increasing influence of Catholics and conservative Protestants. Scholars have examined various indicators of disproportionate influence of one religious group over another, such as looking at the religious affiliation of those in *Who's Who* over time, or professional societies, or members of Congress, or government elites (Davidson et al. 1995; Davidson 1994), or who is getting college degrees (Wuthnow 1988), and the story is the same: the elites in American society are increasingly looking like the rest of the society, religiously speaking. Whereas half a century ago most members of the elite were Episcopalians, Congregationalists, Presbyterians, or Unitarians (e.g. mainline Protestants), now many are Southern Baptists, Adventists, Nazarenes, Pentecostals (e.g. evangelicals), Jews, and Catholics. It was probably only a matter of time before the more conservative religious traditions demanded that their views be heard at the elite table.

What is important for present purposes is that "progressive bioethics" is the liberal wing of the elite culture war and the inheritor of the institutionalized mainstream bioethics of the past 40 years. To put it differently, bioethical debate over the past 40 years has been "progressive"; it just wasn't necessary to label it as such until its explicit opposite came on the scene. And, to get to the heart of the matter, "progressive bioethics" has therefore inherited the orientation to religion of mainstream institutional bioethics, which is to eschew any explicit religious discourse *as well as* any debate about the deeper issues that theologians want to talk about. Moreover, it at least implicitly defines itself as "not religious," in contrast to its opponent, "conservative bioethics." This is evident in the descriptions of the "culture war" in bioethics. Arthur Caplan describes the culture war as concerning "what role ideology and religion ought to play in determining the policies and practices of biomedicine in the world's most powerful state." He writes: "On one side is an alliance of neoconservative and religiously oriented bioethicists" who "speak in terms that are religious or quasi-religious. ... On the other side stand a loose amalgam of left-liberal bioethicists tenuously allied with a far smaller number of more libertarian bioethicists [who] speak primarily in secular terms drawn more from philosophy or the law. Explicitly religious arguments get them nervous." (2005, 12) We need to ask whether it is necessary for progressive bioethics to eschew religious perspectives, even if translated to "quasi-religious" secular language.

Challenges and Opportunities for Progressive Bioethics and Religion

Let us start by being clear about which "religions" we are talking about here. Progressive and conservative bioethical debates are primarily elite phenomena that, while perhaps attempting to motivate or speak to the average person, involve "action" only at the elite level. The "religious" person who is trying to be integrated into these versions of bioethics is, then, also a member of an elite—perhaps a theologian, a clergy member, or a denominational official. It is important to note that on the issues with which bioethics is concerned the average religious person is in the middle of the road, whereas the religious elites take extreme positions. That is, while the average evangelical is more conservative than the average mainline Protestant in regard to abortion, both groups of average people are in the "mushy middle" (Evans 2002b), while evangelical elites are nearly uniformly opposed to abortion and mainline elites are disproportionately pro-choice. Therefore, owing to what I perceive to be a number of ethical conclusions built into progressive bioethics (e.g., a non-absolutist approach to the death of embryos), I think it is safe to say that evangelical and traditionalist Catholic elites are not targets for outreach from the progressive camp. The "challenge" of which I speak is whether or not the *liberal* religious elites can or should join the progressive bioethics cause. To put this in a more historical framework, should liberal religious elites re-join the mainstream institutional bioethics debate, now represented by progressive bioethics, which they left many years ago?

Progressive bioethics, the inheritor of the mainstream public bioethics tradition, will continue to require that religiously oriented people translate their explicitly theological claims into a neutral ethical language. For example, progressive bioethics will not tolerate people justifying arguments with scriptural warrants or knowledge of divine will. This is due in part to progressives' concern with pluralism and diversity, but in part it is because analytic philosophy is much more strongly represented on the progressive side of the divide than on the conservative side. Analytic philosophy in many ways defines itself as *not* being religion, or not making arguments from a religious perspective, but instead using logic and reason derived from a secular stance. This is not to impute malice; it is just that in my opinion analytic philosophers are rarely swayed by religiously based arguments. Caplan generously comments that religious arguments make them "nervous" (2005, 12).

However, this translation requirement does not exclude all religious voices. Some liberal religious elites are willing to engage in this translation, and even see it as a requirement of the faith. For example, the Catholic theologian Lisa Sowle Cahill sees a divide on this issue among theologians. She writes that the consensus for at least two decades concerning the role of theology in public affairs has included the idea that "the only appropriate 'public' language in which to justify, qualify, and reconcile liberties and rights is neutral, secular, and rational" (Cahill 2006, 37). This has divided the theological community. On the one side, "many theologians have tried to adopt the terms of the public language when arguing about public matters, for example, abortion, infertility therapies, euthanasia, or health-care reform." On the other side, "a significant number of others urge fellow theologians to stay away from 'secular bioethics' for two reasons. The first is that persons of faith would never be able to advocate for their values and have an impact in the secular public realm. The second is that the very attempt to do so dilutes distinctly religious values and identities, so that they are disabled from producing religiously authentic behavior on biomedical issues, even with faith communities." (ibid., 37)

So, in sum, outreach from progressive bioethics will be to the segment of the liberal religious community that is willing to translate their concerns into a secular language. I think this is a considerable portion or a majority of this religious community. There is, however, one final restriction that could be put upon religious people trying to engage in progressive bioethics. To my mind, the decision as to whether or not to place this restriction at all will be determinative as to whether a significant number of liberal religious people participate in progressive bioethics. The restriction is that they leave some of their concerns behind: the debates about ends (as I have called them elsewhere) (Evans 2002a), the "thick" concerns, the "big questions," even when they use secular language. Again, it is not necessary that these concerns be left behind. Whereas I have argued that progressive bioethics is the inheritor of the mainstream bioethics tradition that generally has eschewed such questions, times have changed. Most notably, progressive bioethics is having to define itself relative to the conservative, saying "what we believe in." So, these big questions are necessary. Moreover, in Evans 2002a I argued that the avoidance of the "big questions" was due to the need of the bioethics debate to speak to the bureaucratic state, such as government ethics commissions and government agencies. However, to the extent that all of the bioethics issues have become "politicized"—that is, resolved through normal politics involving our elected officials—this requirement lessens. If the ethics of cloning are going to be hammered out in the House of Representatives instead of by a government expert panel at NIH, then the big questions will be more welcome.

So, the focus on the "thin" debate about maximizing taken-for-granted ends may actually be institutionalized detritus from another era. Indeed, as some have argued, the progressives have no chance of winning on their issues with the public unless they engage in the deep substantive questions (Hart 2001). In the remainder of this chapter I will discuss the tensions that will exist between progressive bioethics and the liberal religious community if the debate about ends *is* encouraged and those that will exist if it *is not* encouraged.

If Big Questions Are Not Allowed into the Debate

Let me immediately add some nuance. These two scenarios are hypothetical endpoints. At present, it is not that there are *no* big questions allowed into mainstream bioethical debate as practiced by progressives. Rather, there is a tendency to de-emphasize those questions. I would argue that it has been the remaining religious voices within institutionalized mainstream bioethics that have been pushing these big questions as far as they can without pushing themselves over the precipice that is the boundary of acceptable discourse. I think the question is actually whether this newfound bifurcation of bioethical debate into conservative and progressive camps will lead to a purification drive whereby all of the religious input will be assigned to the conservative camp, all of the "big questions" will be abandoned to this group, and the progressives will move on to promoting the existing ends they have institutionalized, such as the curing of disease and the promotion of scientific discovery. It has never been easy to adjudicate the perspectives of those religious people who do participate in public bioethics. They could easily be defined out of progressive bioethics in order to make debates easier to manage.

If progressive bioethics merely tolerates this tendency of the religious to ask the deeper questions, instead of celebrating it, we should then ask why progressive bioethics would want to attract the religious to its camp. One answer is that the religious are a majority of the population of the United States, but this seems to be an argument for not scaring away the religious, as opposed to incorporating them into the debate in such a way that their religiousness comes through. If that is the concern, I do not think there is much to worry about, because all of the religious people in the privatized religious culture of the United States have a "secular side" that they use to make all sorts of decisions. That would include the bioethical.

We could also imagine an argument that religious voices in progressive bioethics could be useful as either political legitimation or an avenue for mobilization. Both are not worth the effort, in my opinion. First, most people in the United States are aware that influential, serious and devout religious leaders in the U.S. differ on almost every issue described as "bioethical." Thus, having people identified as "religious" taking a stance does not help discredit your opponents who claim that God is on their side. Second, the political mobilization structures of the denominations that might side with progressive bioethics—primarily the mainline Protestants and the more liberal Jewish movements—will not be of much use. The mainliners, whose mobilization ability is not very strong to begin with, will by and large actively stay away from mobilizing political activity on these issues because they will be divisive within their traditions (Evans, forthcoming). The Jewish movement's organizing structures are more efficient, but ultimately do not speak to that many people and are by and large focused on issues perceived of being of greater priority to the Jewish community.

What, then, are the advantages to discouraging religious voices from entering into progressive bioethics? Most notably, progressive bioethics gets to keep the advantages that have accrued to traditional mainstream bioethics by not having debates about which ends to pursue (Evans 2006). People who work in Executive Branch agencies do not want to have debates about what the proper goals of society should be. That would make it appear as if they were setting the values for the society. They want to implement what they can construe as the values of the society (e.g., autonomy and beneficence) in the most efficacious manner. Hence the appeal of mainstream bioethics: the discipline is perfectly situated to answer the question of whether technology X will or will not violate these taken-for-granted ends of society. It does not engage in debate about what the ends of society should be.

To the extent to which the decisions of real consequence in bioethics are made by people other than the public and its elected officials, there is an advantage in not having debates about ends (Evans 2002a). For example, if decisions are going to be made by a panel at the National Institutes of Health, or by a committee at a professional association, then retaining a thinner debate will be beneficial, and this thinner debate requires excluding people who want to talk about the "big" issues. Whether or not the real action in bioethics will remain at a distance from the public is, however, an open question that I will discuss more below.

Ultimately, unless the big issues are explicitly allowed into progressive bioethics, and not just tolerated on the margins, as I would argue is the current situation, there is no reason to try to bring in the liberal religious voices. Yes, many progressive bioethicists are personally religious, but if that is not evident from their bioethical work—if they put on their secular hat when heading to the office—then they are not "religious" for the purposes of progressive bioethics. If people cannot participate as religious people, but must purely use secular reasoning, then we are left with a situation in which a "religious ethicist" in debates over science and medicine is "a former theologian who does not have the professional credentials of a moral philosopher" (Gustafson 1978, 386).

If Big Questions Are Allowed into the Debate

The other path would be to say that progressive bioethics wants to create a space for a debate about the big questions that the religious liberals are disproportionately interested in, and wants to develop a mechanism for this debate to influence the thinner, more pragmatic part of progressive bioethics. This would, I think, bring this religious population into progressive bioethics. That would have certain advantages. Gustafson describes four types of moral discourse: ethical, prophetic, narrative, and policy. For present purposes, I will focus on ethical and prophetic discourse.

Ethical discourse is akin to the debate I have been calling "thin"—that is, the debate about which acts will forward established and institutionalized ends. "How one ought to act in particular circumstances" is Gustafson's definition (1990, 129). He thinks this discourse is critical, and it is indeed the focus of progressive bioethics. While ethical discourse is "micro," prophetic discourse is "macro." In one of these prophetic forms, indictment, the individual is "not occupied with surface issues but expose[s] the roots of what is perceived to be fundamentally and systematically wrong. Prophets are seldom interested in specific acts except insofar as they signify a larger and deeper evil or danger." (Gustafson 1990, 130) This is the debate about ends that, I have argued, the liberal theologians are more interested in than the ethical discourse.

The reason to have prophetic discourse in the discussion is, according to Gustafson, that while prophetic discourse "often looks global and unrealistic to the policy maker ... its perspective can function to jar institutions from blind acceptance of the status quo. . . . To focus moral discourse about medicine too exclusively on what I have described as ethical tends to lose sight of realms of choice and activity that are of great importance." (ibid., 141)

I think that it is useful to have a debate about the ends we are to pursue, to keep us from "blind acceptance of the status quo." Members of a community defined as "progressive" should be asking what ends should be pursued. These discussions should not only happen before the writing of a manifesto, but should happen continuously. When debating the structure of the U.S. Constitution, Thomas Jefferson argued that we should have a revolution every 19 years and re-write the constitution, so that the hand of the dead does not weigh too heavily on the shoulders of the living.[5] Having a space for this sort of debate about ends is necessary, and it is the sort of activity that the secular speaking theologians would gravitate to.

But I think that the current leaders of progressive bioethics need to think about whether this is what they truly want. As an example of what it might mean to truly invite religious liberals into the discussion for a debate about ends—about the deeper questions—let me briefly give an example of what such a debate might look like. If there is any "deep question" that seems to animate religious liberals who write at the margins of bioethical debates, it is the issue of justice. Indeed, Cahill sees that the primary purpose of theology in public bioethics is to focus on this issue of justice. She writes that theology is "an indispensable conversation partner in the realm of public bioethics, for any society that aims to incorporate health care and health research within institutions and practices that serve justice and the common good" (2006, 55). Clearly, Cahill sees theology as the voice for justice. Progressives might think that this would be easy to incorporate, in that "justice" is one of the ends we are to pursue in the principlist model of bioethics that is the most influential model on the progressive side.

But the religious liberals will want to debate what "justice" really means. And it is clear that they do not think it means what mainstream institutional bioethics has taken it to mean. In contrast to the mainstream institutional bioethics tradition that sees justice as treating all people equally, some liberal theologians see it as mandating a preference for the poor over the wealthy. For example, the liberal Protestant theologian Karen Lebacqz writes:

> While mainstream bioethics still tends to relegate justice to questions of allocation of scarce resources or questions of access to new technologies, the biblical image of *sedakah* suggests that a major overhaul of the system might be needed. If justice requires not simply treating equals equally, but treating the marginalized and oppressed with special consideration, then some approaches to health care should get priority. For example, preventive medicine, which gets very little attention in bioethics today, might emerge as the preferred option. Basic nutrition and sanitation might be deemed more important than the building of neonatal intensive care units. Further, if the epistemological privilege of the oppressed is taken seriously the perspective from which biomedical practices are assessed would have to change. (2006, 259)

Cahill has a similar view, but from the Catholic tradition. She writes that it is "the theological agenda to seek justice in health care is defined generally by the biblical 'option for the poor.' Specifically, one of its most important and distinctive contributions to public discourse is a critique of the ways in which modern biomedicine and biotechnology have become luxury items marketed to economically privileged classes, while the world's poor majority lacks basic health needs." (2006, 55)

From all of this we may begin to see why progressive bioethics may not want to invite these people into the conversation, but rather may want to limit them to using the terms of the debate developed in mainstream institutional bioethics—the "ethical" language, in Gustafson's terms. The university medical center that houses the bioethics center may not take kindly to the idea that embryonic stem cell research should end so that money can be spent on basic health care for the poor in Africa, which to my mind would be an implication of what Lebacqz and Cahill are claiming. Lebacqz gives another example: "Even so fundamental a right as the right to hold property may give way to the demands of justice seen in the needs of others. For example, *Gaudium et Spes* reports the consensus of Vatican II that a person in extreme necessity may take from the riches of others to meet basic needs. Thus, in the dominant Roman Catholic tradition, the demands of justice can take priority over individual desires, autonomy, and rights." (2006, 258)

To reiterate, serving the needs of the oppressed should take precedence over private property. This may not be what the board of directors wants to hear. What liberal Protestants (along with Gustafson) would call "prophetic" discourse will probably be called in the mainstream bioethics world "unrealistic" or "utopian" discourse. It is a focus on what ends to pursue, the "big questions." Claims about these big questions are not precise policy proposals, which mainstream institutional bioethical debate prefers. In my experience, this is what mainstream bioethics professionals hate about liberal religious discourse. It is most definitely not a micro debate about whether medical practice X will produce more justice than medical practice Y. The religious voices want to define the criteria for what we should be pursuing, and let the details be left to others.

Conclusion

The history between bioethics and religion structures the options for progressive bioethics' encounter with religion. Any encounter will have to be with those theologians who are willing to translate their concerns into secular terms, which is in my estimation a fairly sizable proportion of the liberal religious group. However, if progressive bioethics is not ultimately interested in the "big questions" it may be harder to get this group interested. This, of course, makes one wonder why people in progressive bioethics want religion to be involved anyway. Of course, many progressive bioethicists are personally religious, but there is already a mechanism in American society in place for dealing with this issue: religious privatization. Religion is limited to one's personal life, with boundaries drawn between it and other aspects of one's life. This is not a criticism of liberal religious people. Indeed, it could be argued that liberal religious people invented the idea of religious privatization in order to fulfill their own theological ideals (Demerath 1995). Of the large religious groups in the United States, it is today the mainline Protestants who are the most privatized, at least when it comes to interacting with the public sphere (Regnerus and Smith 1998).

This seems a time for progressive bioethics to be thinking about what it stands for, or, in my language, what its ends are. Indeed, this volume seems to be a first attempt at that. At a minimum, this is a conversation that religious liberals would not only want to have, but would make a positive contribution to.

Would it be naive of me to think that the culture war in bioethics can be thought of as a trial separation and not a divorce? If there is to be reconciliation, both sides will have to give in a bit, or at least be able to speak to each other. The neo-conservative religiously oriented bioethics (by which is really meant *conservative* religious bioethics) is more focused on the "big questions" than is the progressive side. This difference is clear from observing the President's Council on Bioethics, which often focuses on these sorts of issues. If there is to be a rapprochement, it will require finding some shared ground between the two. If progressive bioethics can retain its liberal religious people, with their focus on the big questions, there will be at least a mediating faction between the two sides that could re-unite the previous, if somewhat troubled, marriage.

Notes

1. I make a distinction between "bioethics," which is a profession like theology, sociology or biology, and "bioethical debate," which is a public debate among

elites about what should be done about medicine and technology, a debate that people from any profession could contribute to. See Evans 2002a, 34.

2. In my experience, people in elite bioethics debates do not like to be called "elite" because it denotes "elitism." I use the term in its more sociological usage, which is as "a minority group which has power or influence over others and is recognized as being in some way superior." (See Abercrombie et al. 1988, 84.) This is the point of elite bioethical debate: that people in the debate know more than the average person and are therefore more qualified to engage in the debate.

3. I would say that Paul Ramsey would be the possible exception to this generalization.

4. Most people credit Jimmy Carter as the first evangelical president. In my view, Carter, while a born-again Southern Baptist, was from an earlier generation of particularly Southern evangelicals who were actually more mainline in their orientation to the broader world. Put differently, I think that the evangelicalism that George W. Bush converted to in the mid 1980s was different than the evangelicalism of Carter's formative years. On Bush's conversion, see Suskind 2004.

5. Letter, Jefferson to James Madison, September 6, 1789 (Appleby and Ball 1999).

References

Abercrombie, Nicholas, Stephen Hill, and Bryan S. Turner. 1988. *The Penguin Dictionary of Sociology*, second edition. Penguin Books.

Appleby, Joyce, and Terrance Ball. 1999. *Thomas Jefferson: Political Writings*. Cambridge University Press.

Cahill, Lisa Sowle. 2006. Theology's role in public bioethics. In *Handbook of Bioethics and Religion*, ed. D. Guinn. Oxford University Press.

Caplan, Arthur. 2005. "Who lost China?" A foreshadowing of today's ideological disputes in bioethics. *Hastings Center Report* 35, no. 3: 12–13.

Davidson, James D. 1994. Religion among America's elite: Persistence and change in the Protestant establishment. *Sociology of Religion* 55: 419–440.

Davidson, James D., Ralph E. Pyle, and David V. Reyes. 1995. Persistence and change in the Protestant establishment. *Social Forces* 74: 157–175.

Demerath, N. J. 1995. Cultural victory and organizational defeat in the paradoxical decline of liberal Protestantism. *Journal for the Scientific Study of Religion* 34: 458–469.

Engelhardt, H. Tristram. 1986. *The Foundations of Bioethics*. Oxford University Press.

Evans, John H. 2000. A sociological account of the growth of principlism. *Hastings Center Report* 30, no. 5: 31–38.

Evans, John H. 2002a. *Playing God? Human Genetic Engineering and the Rationalization of Public Bioethical Debate.* University of Chicago Press.

Evans, John H. 2002b. Polarization in abortion attitudes in U.S. religious traditions, 1972–1998. *Sociological Forum* 17: 397–422.

Evans, John H. 2006. Between technocracy and democratic legitimation: A proposed compromise position for common morality public bioethics. *Journal of Medicine and Philosophy* 31: 213–234.

Evans, John H. Forthcoming. Where is the counter-weight? Explorations on the decline of mainline Protestants in public debates. In *The Christian Conservative Movement and American Democracy*, ed. S. Brint and J. Schroedel. Russell Sage Foundation Press.

Gustafson, James M. 1978. Theology confronts technology and the life sciences. *Commonweal* 105: 386–392.

Gustafson, James M. 1990. Moral discourse about medicine: A variety of forms. *Journal of Medicine and Philosophy* 15: 125–142.

Hart, Stephen. 2001. *Cultural Dilemmas of Progressive Politics: Styles of Engagement Among Grassroots Activists.* University of Chicago Press.

Jonsen, Albert R. 2006. A history of religion and bioethics. In *Handbook of Bioethics and Religion*, ed. D. Guinn. Oxford University Press.

Lebacqz, Karen. 2006. Philosophy, theology, and the claims of justice. In *Handbook of Bioethics and Religion*, ed. D. Guinn. Oxford University Press.

Lysaught, M. Therese. 2006. And power corrupts: Religion and the disciplinary matrix of bioethics. In *Handbook of Bioethics and Religion*, ed. D. Guinn. Oxford University Press.

Noll, Mark A. 1992. *A History of Christianity in the United States and Canada.* Eerdmans.

Regnerus, Mark D., and Christian Smith. 1998. Selective deprivatization among American religious traditions: The reversal of the great reversal. *Social Forces* 76: 1347–1372.

Smith, Christian. 1998. *American Evangelicalism: Embattled and Thriving.* University of Chicago Press.

Suskind, Ron. 2004. What makes Bush's presidency so radical—even to some Republicans—is his preternatural, faith-infused certainty in uncertain times. *New York Times Magazine*, October 17.

U.S. Congress, Office of Technology Assessment. 1993. *Biomedical Ethics in U.S. Public Policy—Background Paper.* Government Printing Office.

Wuthnow, Robert. 1988. *The Restructuring of American religion.* Princeton University Press.

Wuthnow, Robert, and John H. Evans. 2002. *The Quiet Hand of God: Faith-Based Activism and the Public Role of Mainline Protestantism.* University of California Press.

8

Can National Bioethics Commissions Be Progressive? Should They?

Eric M. Meslin

On January 29, 2000, I found myself sitting in a taxi with Richard Doerflinger of the National Conference of Catholic Bishops for the hour-long trip from Dartmouth College to the airport in Manchester, New Hampshire. Doerflinger and I had just spent the day at a meeting organized by Ron Green, Director of the Ethics Institute at Dartmouth, on the subject of embryonic stem cell research, and since we were both returning to Washington, Green organized a shared taxi.

The thought of spending an hour with Doerflinger was somewhat unnerving. Not only were our views about the ethics of embryonic stem cell research diametrically opposed on almost every feature of this debate, but he had also proved himself to be a formidable advocate for his position, whether it was testifying before House and Senate subcommittees, speaking at meetings sponsored by the National Institutes of Health, or giving lectures at professional conferences. And, of course he had twice appeared before the National Bioethics Advisory Commission as it prepared its report on the stem cell issue (National Bioethics Advisory Commission 1999). As the NBAC's Executive Director during that time, I knew Doerflinger to be well read on the science, ideologically committed to his beliefs, and a staunch defender of his position. This potent combination is exactly what is needed when advisory commissions seek input on controversial topics, but when the person is sitting three feet away from you in the back of a cab it can be unsettling. What were we supposed to talk about?

It was something of a surprise, therefore, when, after a few uncomfortable moments of small talk, Doerflinger turned to me and said (I'm paraphrasing) "Eric, I've got to tell you, I don't agree with any of the NBAC's recommendations on funding for embryonic stem cell research, but I respect the commission's ethical consistency." The consistency he was referring to was the NBAC's conclusion that since no ethically relevant line could sensibly be drawn between the derivation of embryonic stem cells from early human embryos (an extraction process that destroyed the embryos) and the use of these cells for scientific research, then no such line should be drawn for purposes of federal funding. Doerflinger's statement, or at least my memory of it, has stuck with me ever since. I took him to be making a point as poignant as it was pragmatic: the NBAC had gone much too far for his liking in recommending that the NIH should be able to fund research that both derived and used embryonic stem cells, but he understood the reasoning behind the recommendations and had a certain amount of professional appreciation for the effort.

What Doerflinger didn't say—and I'm not sure he would have put it this way—was that the NBAC's position was progressive in at least two senses. First, its position on embryonic stem cell research was a pragmatic policy recommendation based on sound science and sound ethics. It provided a clear analysis of challenges that arose from a new development in science and technology and offered specific ways to resolve it, some of which were policy and regulatory changes, some of which were changes in procedures, and some of which were calls for action by those outside government (National Bioethics Advisory Commission 1999). Second, it was progressive in that the NBAC's recommendations were not burdened by myopic ideological thinking. The NBAC did not begin with a presumption about (or narrow definitions of) the sanctity of life, human dignity, or the absolute moral status of the embryo and then try to work toward a set of recommendations that cohered with these prior beliefs and values. That's not to say the recommendations were value free, or that they were devoid of careful ethical reflection. On the contrary, substantive ethical analysis and deliberation was used throughout the commission's work on this issue. Rather, the commission took this view:

Conscientious individuals have come to different conclusions regarding both public policy and private actions in the area of stem cell research. Their differing perspectives cannot easily be bridged by any single public policy. But the development of public policy in a morally contested area is not a novel challenge for a pluralistic democracy such as that which exists in the United States. We are profoundly aware of the diverse and strongly held views on the subject of this report and have wrestled with the implications of these different views at each of our meetings devoted to this topic. Our aim throughout these deliberations has been to formulate a set of recommendations that fully reflects widely shared views and that, in our view, would serve the best interests of society. (National Bioethics Advisory Commission 1999, iii)

Doerflinger's comments may have unintentionally given tacit acknowledgement to the NBAC's progressive activity, but history records that the NBAC's pragmatic approach did not win the day. In the previous month (December 1999), the NIH had issued draft guidelines on embryonic stem cell research that would permit funding only for the *use* of embryonic stem cells and not for their derivation—a policy that had been expected given that it was consistent with the rider that had been placed on every Department of Health and Human Services (DHHS) appropriations bill since 1996 prohibiting research that created or destroyed human embryos. Indeed, in a memo dated January 15, 1999, Harriet Rabb, the DHHS's general counsel at the time, had advised NIH Director Harold Varmus that "the statutory position on the use of funds appropriated to HHS for human embryo research *would not apply* to research utilizing human pluripotent stem cells because such cells are not a human embryo within the statutory definition." Varmus presented Rabb's legal opinion when he appeared before the NBAC on January 19, 1999 (National Bioethics Advisory Commission 1999, 34). From their point of view, the NIH's position was just as pragmatic, in that it permitted valuable research without having to re-open the Dickey amendment, which prohibits federal funding for research that creates or destroys embryos.

The Dartmouth cab ride, the Rabb memo, and Dr. Varmus's NBAC appearance are small vignettes from the discussion of stem cell research policy more than a decade ago, just one of the many science policy debates that are played out inside the beltway in Washington on a regular basis. Stem cell research wasn't the first politically charged topic that the NBAC faced—cloning was the first, after the birth of Dolly in July 1997 and President Clinton's request to the NBAC to "undertake a thorough review of the legal and ethical issues . . . and report back to me in ninety days" (National Bioethics Advisory Commission 1997). Nor was the NBAC the first U.S. bioethics committee to engage ethical issues in the life sciences (U.S. Congress, OTA 1993). Federal advisory committees focusing on bioethics topics have been established or have functioned off and on under every administration since Richard Nixon's. While the Tuskegee Syphilis Study Ad Hoc Panel may have been one of the first to focus on such issues (Tuskegee 1973; Jones 2008), it was the National Commission for the Protection of Human Subjects of Biomedical and Behavioral Research (1974–1978) that began the current era of government advisory committees established to provide advice on bioethics.

In the three decades since, many U.S. bioethics commissions have often acted as progressive bodies—pragmatic, policy focused, and ideologically neutral—though it is not clear that this was always their intention. For the most part, bioethics commissions have been used as topic-limited and time-limited bodies that concentrate on major life science issues. They tend not to focus on science policy challenges that occur higher up the policy food chain, issues that may have their origins in other areas of domestic policy (e.g., labor, commerce, environment) and foreign policy, including national security. Moreover, they function as advisory committees to government, without the power to implement their findings. It is one thing to recommend a sensible, pragmatic approach to an ethical problem when the authority to accept or reject rests with others. This chapter will describe how recent commissions may be seen as progressive, and consider what may be gained or lost were we to more intentionally adopt a progressive model for domestic bioethics commissions.

Science Policy Advice as Progressive

The reliance on experts to inquire into matters of social and political importance has been a feature of civil society for more than two millennia[1]: from the Great Library at Alexandria, which promoted intellectual debate, to groups like the Compagnie du Gai Sçavoir, the Academia Platonica, and the Barber Surgeons of Edinburgh (later the Royal College of Surgeons of Edinburgh), scientific expertise was sought and used by the first estate. But it took until the seventeenth century for the British

Royal Society and the French Academy of Sciences, and other scholarly societies to begin to conduct inquiries as a service for the government. Later, as the relationship between matters of science and the needs of government grew stronger, government-commissioned advisory bodies began to supplement (if not replace) the advice provided by these independent scholarly societies. Specialized bodies, such as Royal Commissions and "blue ribbon commissions," became more common throughout the eighteenth and nineteenth centuries, and were appointed for the sole purpose of reporting on topics of importance to the monarchy.

In the United States, the scientific establishment was initially resistant to the idea that science and politics could co-exist on equal terms. Scientists wanted to maintain a certain level of independence from government control and resisted the development of any state-sponsored institutions for research (Greenberg 1967). By the middle of the twentieth century, commissions no longer consisted of politicians and political advisors, and included civilians with diverse expertise in relevant academic disciplines. This move contributed to a democratization of science and research, and had a profound influence on the nature of the relationship between government, science, and society. So it is not surprising to see bioethics commissions emerging in the United States (and elsewhere) as instruments of public policy construction.

By 2005 bioethics commissions existed in at least 85 countries, on every continent except Antarctica (Meslin and Johnson 2008). For many countries, national commissions are one of the first sources of consultation for their respective governments on emerging issues in science and technology (Shapiro and Meslin 2005). Perhaps this is because bioethics commissions occupy a particular niche in the ongoing public conversation about health and science policy. Their visibility, convening power, and proximity to senior government officials provide uncommon opportunities to speak about some of society's most pressing issues. Nevertheless, these very factors, and the particular niche that commissions occupy within the federal bureaucracy, may also constrain their ability to function as a progressive organization or to advance a progressive agenda. In an early authoritative report on U.S. bioethics commissions, the now-defunct Office of Technology Assessment summarized the recognized strengths of government commissions as follows:

National commissions provide a vehicle to handle issues that are amenable to consensus building, or at least to an elaboration of conflicting views. Ideally, they garner the esteem of policymakers and experts by serving as a forum to: crystallize a consensus or delineate points of disagreement; identify emerging issues; defuse controversy or delay decision making; propose regulations, develop guidelines, or formulate policy options; review implementation of existing law and policies; aid judicial decisionmaking; educate professionals and the public; and promote interdisciplinary research. (U.S. Congress, OTA 199327)

To varying degrees all bioethics commissions carry out these functions. The National Commission, the President's Commission for the Study of Ethical Problems in Medicine and Biomedical and Behavioral Research (President's Commission), the Advisory Committee on Human Radiation Experiments, the NBAC and the recent President's Council on Bioethics are all part of the legacy of U.S. national bioethics commissions. Common to all is a focus on particular topics (research involving human subjects) or groups of topics (genetics, health care, biotechnology)—topics found in the canon of bioethics, which includes the great cases and controversies from the past four decades: research involving human subjects, access to health care, gene patenting, end-of-life decisions, genetic studies using tissue banks, international clinical trials, mental health research, the limits of biotechnology, caring for the aging population, and medical futility. These topics fill anthologies and journals, occupy blogs, attract media attention, form the basis of research programs, and become the focus of advocacy organizations, institutional policies, state and federal legislation, and international harmonization efforts.

All U.S. bioethics commissions have made contributions to the public conversation on these topics and have had a demonstrable impact on policy (Meslin 2003; Eiseman 2003). And yet the reason many cases and controversies remain a staple for commission deliberation is the fact that they are never fully resolved. The National Commission's work has been widely praised, particularly due to the long-standing influence and value of the Belmont Report, but there continues to be a robust discussion about the depth, applicability, and relevance of these principles to current research policy (Childress et al. 2005). Several other reports, including those on the fetus (National Commission for the Protection of Human Subjects of Biomedical and Behavioral Research 1975), prisoners (National Commission for the Protection of Human Subjects of

Biomedical and Behavioral Research 1976), and children (National Commission for the Protection of Human Subjects of Biomedical and Behavioral Research 1977), had direct and specific influence on the development of federal research regulations, but these topics have not disappeared from the national stage. The President's Commission's reports on defining death (1981) and forgoing end-of-life treatments (1983) led to significant changes in regulatory regimes and professional standards of the day, whereas others on topics such as compensation for injured research subjects (President's Commission for the Study of Ethical Problems in Medicine and Biomedical and Behavioral Research 1982), securing access to health care (President's Commission for the Study of Ethical Problems in Medicine and Biomedical and Behavioral Research 1983b), and protecting human subjects (President's Commission for the Study of Ethical Problems in Medicine and Biomedical and Behavioral Research 1981, 1983a) provided strong arguments for implementing policy, though we are still awaiting government action. Indeed, it took a full decade before the President's Commission's recommendation that there be a common set of human subjects regulations in the United States were substantially implemented via the Federal Policy for the Protection of Human Subjects—and even then only 16 of the more than 50 federal agencies agreed to be signatories to the "Common Rule."

The unique contributions of the Advisory Committee on Human Radiation Experiments (ACHRE) flowed from its particular role in investigating and reporting on the use of human subjects in federally sponsored (mostly military) research using ionizing radiation. Not only did this commission undertake and complete its assigned task in a landmark report (ACHRE 1995), it left a historical record of its work that will benefit the public and scholars for decades by declassifying millions of pages of previously classified government data.

The truth is that science, and by extension science policy, is intensely political, partisan, and messy. It is hard to separate the ethics from the politics, and the politics from the policy. We have known this at least since Daniel Greenberg concluded his book *The Politics of Pure Science* with the following quote from John F. Kennedy: "Scientists alone can establish the objectives of their research, but society, in extending support to science, must take account of its own needs." Greenberg (1967, 292) called Kennedy's words "the supreme characterization of the predicament that underlies the politics of pure science."

Greenberg's observation illustrates the difficulty in assessing the impact of advisory commissions generally (Eiseman 2003), though perhaps the impact is not as minor as Berger and Moreno suggest in their chapter in this book. For example, when considering the impact of the ACHRE, Ruth Faden suggests that commissions should be judged by the "degree to which the president supports the commission's findings and recommendations," a criterion suggested by noted presidential advisory commission scholar Thomas Wolanin (Faden 1996, 9). By this criterion, the ACHRE was very successful, and the President's Commission and the NBAC were somewhat successful.

But committees also have impact in ways besides the outcomes of their deliberations. They can have an impact on the way that deliberations occur. Consider three examples. First, the FACA requirements for ensuring public accountability were constructed in the 1970s, well before the Internet. Access to meeting information, draft reports, transcripts, and the content of briefing books provided the public with increased access to, and therefore greater participation in, the commission's activity. Coupled with expert and public testimony at meetings, solicited comments on commission draft reports have had a profound impact on commission deliberations (Meslin 1999). It is now common practice to use the Web to inform and engage the public—an important feature of the progressive model.

A very different example of the impact of a commission arose from the experience of the President's Council on Bioethics (PCB) when two of its members—Elizabeth Blackburn and Bill May—were not reappointed by President George W. Bush. Blackburn's departure from the PCB was especially noteworthy due to the widely held view that she was not re-appointed because her defense of embryonic stem cell research was at odds with the policies of President Bush (Meslin 2004). The Blackburn/May story raised this question: How does membership influence the work of these bodies? Given that the progressive model for policy development encourages public engagement, commission membership will always stand as a case study in stakeholder relations.

Consider the histories of several commissions as examples of the way in which commissions can be used as instruments of policy:

• In 1978–80, the Ethics Advisory Board was established to review protocols involving *in vitro* fertilization research, and yet it was effectively disbanded when its budget was eliminated by President George H. W. Bush.

• In 1985, Congress established the Biomedical Ethics Advisory Committee, overseen by a Bioethics Advisory Board, in an effort to develop a bipartisan bioethics committee separate from the Executive Branch. Yet it met only once in 4 years because of difficulties in agreeing about the stance that its members should take on issues such as abortion (U.S. Congress, OTA 1993).

• In 1988, the NIH convened a Human Fetal Tissue Transplantation Research Panel, the recommendations of which would have created guidance for permitting research while protecting women against potential coercion or exploitation. Despite widespread support, President Ronald Reagan issued an executive order prohibiting federal support for fetal tissue transplantation research (an order that was extended by President George H. W. Bush in 1989, lifted by President Bill Clinton in 1993, and reinstituted by President George W. Bush in 2001).

• In 1994, the Human Embryo Research Panel was established to provide advice and recommendations to the NIH on this sensitive topic. Despite its support for this area of study, President Clinton rejected one major recommendation, stating "I do not believe that federal funds should be used to support the creation of human embryos for research purposes, and I have directed that NIH not allocate any resources for such research."

• In 1996, Ruth Faden, chair of the ACHRE, diplomatically described the influence that the ACHRE used in its investigation of human radiation experiments: "The status and authority we enjoyed as a presidential committee were critical to our success in this discovery of the nation's past. The agencies were obligated by executive order to comply with our requests. . . . Moreover . . . we received extensive press coverage, which gave us leverage in our relationships with the agencies." (Faden 1996, 7)

If there was any doubt about which direction NIH policy on embry-onic stem cell research would take, it was confirmed in dramatic fashion on July 14, 1999, when the White House issued an unusual press release indicating its opposition to the NBAC's likely recommendation to support funding the derivation and use of embryonic stem cells, instead favoring NIH's position to only support funding the use of these cells. Later, in an article in the *Kennedy Institute of Ethics Journal*, NBAC chair Harold Shapiro and I wrote about this unexpected White House action:

This statement signaled the White House's opposition to NBAC's draft recom-mendations calling for revisions to existing legislation that would permit federal funding of both derivation and use of embryonic stem cells. The most puzzling aspect of this announcement was its timing: although NBAC had completed deliberations on the report, it had yet to submit it final report to the President. Never before had the White House proactively commented on an unfinished NBAC report. No reason was given for why such precipitous action was neces-sary. (Meslin and Shapiro 2002, 97)

These very different examples of impact share a common thread—all commissions have faced political pressures of one sort or another. Perhaps this is why the Office of Technology Assessment observed that "government-sanctioned commissions allow debates about contentious issues to go forward in a somewhat less politicized way than is possible on the floors of Congress or a State legislature" (27), and explicitly excluded politics from its list of considerations for any future efforts to establish national bioethics committees:

Absent from this list is politics; the very nature of creating a new entity subjects each of these factors to the pressures and whims inherent in the political process. (U.S. Congress, OTA 31)

The Risks and Benefits of Progressive Bioethics Commissions

While commissions are often created because of political necessity, treated to rough-and-tumble political gamesmanship, and encouraged to share their ideas on politically sensitive topics, they are not fully empowered to undertake political action. There is a profound difference between weighing in on issues by crafting reports filled with recommendations— even provocative ones—and acting as a political voice. U.S. bioethics commissions to date have been progressive, though perhaps not intentionally so. They have embodied many of the core progressive principles: "that progress is possible, that pragmatism should prevail over ideology, that both individual rights and the common good can be respected and promoted, and that sound public policymaking requires a respect for evidence and a willingness to change familiar ways of operating" (Berger and Moreno, this volume). They have been transparent, sought public involvement, prioritized the value of scientific rigor in their work, and applied values and principles in the service of solving problems. But insofar as bioethics commissions *in general* may be described as progressive, *individual* commissions have varied in how progressive they were. As microcosms of broader policy deliberation in civic society, subtle differences in power, methods, and emphasis may distinguish one commission's progressivism from another.

The ACHRE's power to investigate radiation experiments, though it did not have subpoena authority, was very different from the NBAC's or the President's Council's. And both the NBAC and the PCB, as Executive Branch committees established by executive order, had different reporting relationships and accountability than the National Commission and the President's Commission, both of which were established by act of Congress.

With respect to methods, some have cautioned that the drive to consensus might signal a "threat to good government ideals of openness and accountability to a diverse public" (Spielman 2003, 355), and much has been made of the PCB's dramatic rejection of the use of consensus. On January 17, 2002, the PCB's chairman, Dr. Leon Kass, spoke clearly and unambiguously about his disdain for consensus in his opening remarks at the first meeting of the PCB:

Next to our manner. If our scope is to be broad our manner of inquiry must be searching and open. We are a diverse and heterogeneous group. By training, we are scientists and physicians, lawyers and social scientists, humanists and theologians. By political leaning, we are liberals and conservatives, republicans, democrats and independents. And by religion, Protestants, Catholics, Jews and perhaps some who are none of the above. I, frankly, have no idea in many cases. But I trust that we share a deep concern for the importance of the issues and the desire to work with people from differing backgrounds in search for truth and wisdom about these vexing matters, eager to develop a comprehensive and deep understanding of the issues. Because reasonable and morally serious people can differ about fundamental matters it is fortunate that we have been liberated from an overriding concern to reach consensus. (Kass 2002)

This debate over the usefulness of consensus may say more about the distinctive challenge for progressive bioethics commissions than any other, since it is not clear whether the PCB's preferred approach of "deep moral engagement" was any more effective than the consensus approaches adopted by other commissions (including the National Commission for the Protection of Human Subjects of Biomedical and Behavioral Research, the President's Commission for the Study of Ethical Problems in Medicine and Biomedical and Behavioral Research, and the National Bioethics Advisory Commission for advancing a pragmatic, ethically defensible science policy agenda. I have argued elsewhere that conducting policy analysis "in public" (my phrase for the manner in which commissions work) requires that people have the opportunity to express their points of view and make judgments about whether and when to allow the interests of the group to take precedence over those of individuals (Meslin 2003, 104). I remain convinced that in the NBAC's case consensus was a method for satisfying this requirement, but it was not an end it itself. The NBAC did not seek this form of negotiated agreement at all costs— which Joel Feinberg (1973) referred to as "the need to please voters, to make deals with colleagues." It sought agreement from commissioners, permitting personal statements from individual commissioners to supplement, qualify, or register concerns about particular recommendations. This approach reflected a principle of deliberation adopted in its discussions about stem cell research: ". . . if it is possible to achieve essentially the same legitimate public goals with a policy that does not offend some citizens' moral sensibilities, it would be better to do so" (National Bioethics Advisory Commission 1999, 57). In this way consensus becomes an ethically defensible instrument of progressivism.

Looking to the Future

It is easy to get trapped in a debate about whether bioethics commissions are or should be progressive by focusing on the past: past commissions, past successes, past rhetoric. Among the features of the progressive approach to policy is the optimistic gaze we ought to cast upon our

future. Setting aside whether the power and methods adopted by commissions are more or less effective for responding to the traditional canon
of bioethical inquiry, we might also consider the topics which commissions might weigh in on that extend beyond the canon.

Should bioethics commissions be sought for advice about the skills,
credentials, and vision of the next NIH director, the next FDA commissioner, or the next DHHS secretary? It's not hard to imagine why this
might be a valuable service: these positions are among the most important in influencing the quality of life of the country. An NIH director's
views on the importance of funding research on embryonic stem cells,
gene therapy, or HIV will influence the appropriations committees that
fund the NIH. Moreover, given the NIH's leading role in assuring standards for responsible conduct of research, wouldn't it be reassuring to
know that the person leading the agency enjoyed the confidence of the
topic bioethics advisors in the country?

The selection of the next FDA commissioner presents similar opportunities for bioethics input. A commissioner who believes that drug safety
is lax and is prepared to slow the pace of marketing new drugs will
function quite differently from a commissioner who believes that drug
safety standards are stifling research and development. The leadership of
this agency, and no doubt others—the Centers for Disease Control and
Prevent, the Center for Medicare and Medicaid Services, the Office of
Science and Technology Policy—will have a direct impact on the health
and welfare of the country. A commission with the authority to recommend (or perhaps provide a grade or a rating) on such nominees would
be a very different one from a commission that advises on the ethical
issues arising from a decision made by one of these leaders. Is the latter
type always more important than the former?

What about the budgets of the NIH, the CDC, and the Public Health
Service? Should a bioethics commission be consulted about the priorities
established in these agencies or the budgets of other federal agencies?
Priority setting and resource allocation have been among the bread and
butter issues of bioethics ever since ventilators were scarce and more
people needed organs than there were organs to transplant. Priority
setting requires thoughtful, real-time deliberation about fundamental
questions concerning justice and fairness; surely an enlightened national

committee of experts and public members could contribute directly to such a discussion.

Is it appropriate for a national bioethics commission to review and recommend the amount of humanitarian aid the United States should provide, to which countries, and under what type of policy rubric? The odious Mexico City Policy, for instance, prohibits private overseas grantees that receive U.S. family planning funding and technical assistance from using their own funds to provide abortions or counsel women about abortions (Cincotta and Crane 2001). Might a national bioethics commission intrude into trade and aid policy? Would it make a difference in the eyes of other countries if the national bioethics commission of the United States held hearings on such a policy, heard witnesses, prepared a report, and recommended revisions to our foreign policy?

It is surely obvious that bioethics, with its accepted focus on ethical, legal, and social issues arising from medicine and the life sciences, has much to say about medicalized torture at Abu Ghraib prison, prisoner abuse at Guantanamo, the Patriot Act's implications for personal privacy, and the justification of the wars in Afghanistan and Iraq. But what if a national bioethics commission had been given authority to review allegations, conduct investigations, recommend action, and require compliance in these situations?

Some might see these questions as naive, or simply wrong-headed. My point is simple: So long as bioethics commissions are seen as topic- and time-limited bodies that focus only on the traditional canon, they will never weigh in on important policy work that occurs higher up the food chain. And so long as they function as advisory committees, their impact will always be qualified. As for whether it is wrongheaded, others can judge whether the risks of altering the structure, function, and voice of a national commission outweigh the benefits.

I have argued that bioethics commissions in the United States have been, more or less, examples of a progressive public policy model, but they have not been intentionally so, even though commissions embody and adopt many of the principles and approaches of progressivism. This is because their structure, function, reporting relationships, and authority have placed some constraints on them. After more than 30 years of commissions whose progressivism has been tempered by these constraints, the time is right to consider how a bioethics commission might be more intentionally progressive. A commission could be established in a very different way—modeled after the Federal Communications Commission, the Nuclear Regulatory Commission, the Federal Reserve, or the Supreme Court (Fletcher 2001). Its topics could extend beyond biomedicine and the life sciences to address broader topics that influence health and economic well being, such as environmental policy, international trade, human rights, and consumer protections. Its membership could change to include more (or fewer) experts and more (or fewer) public members. It could be given investigatory powers, and be established by Congress as a permanent body. Any of these changes are worthy of consideration, as long as they are planned for and evaluated for impact and effectiveness. But what must not change is its capacity for advancing pragmatic, ethically defensible science policy. While it is gratifying to sit in the back of a cab and get the immediate feedback I did from Richard Doerflinger that the NBAC's progressivism was respected, the importance of the future of science policy is too important to rely on these chance interactions.

Acknowledgements

Portions of this chapter are based on remarks I gave at a 2006 meeting titled Bioethics and Politics Past, Present and Future, convened by the Center for American Progress. The chapter was further informed by a 2007 meeting titled Reassessing National Bioethics Commissions in the United States, held at the University of Pennsylvania and sponsored by the Greenwall Foundation. I am also grateful for comments provided on earlier drafts of this chapter by Jonathan D. Moreno, Yuval Levin, Sam Berger, and Peter H. Schwartz, and for editorial assistance from Elizabeth Edenberg.

Note

1. The following two paragraphs are adapted from Meslin and Johnson 2008.

References

Advisory Committee on Human Radiation Experiments. 1996. Final report of the Advisory Committee on Human Radiation Experiments. Oxford University Press.

Berger, Sam, and Jonathan D. Moreno. 2010. Bioethics progressing. In this volume.

Childress, James F., Eric M. Meslin, and Harold T. Shapiro, eds. 2005. *Belmont Revisited: Ethical Principles for Research with Human Subjects*. Georgetown University Press.

Cincotta, Richard P., and Barbara B. Crane. 2001. The Mexico City Policy and U.S. family planning assistance. *Science* 294, no. 5542: 525–526.

Eiseman, Elisa. 2003. *The National Bioethics Advisory Commission: Contributing to Public Policy.* RAND.

Faden, Ruth. 1996. The Advisory Committee on Human Radiation Experiments. *Hastings Center Report* 26, no. 5: 5–10.

Federal Advisory Committee Act. 5 U.S.C. app (1998).

Feinberg, Joel. 1973. *Social Philosophy*. Prentice-Hall.

Fletcher, John C. 2001. Location of the Office for Protection from Research Risks within the National Institutes of Health: Problems of status and independent authority. In *Ethical and policy issues in research involving human participants*, volume II. NBAC.

Greenberg, Daniel S. 1967. *The Politics of Pure Science*. University of Chicago Press.

Jones, James H. 2008. The Tuskegee syphilis experiment. In *The Oxford Textbook of Clinical Research Ethics*, ed. E. Emanuel et al. Oxford University Press.

Kass, Leon. 2002. Welcome and opening remarks. President's Council on Bioethics, January 17, Washington. http://www.bioethics.gov.

Meslin, Eric M. 1996. Adding to the canon: The final report. *Hastings Center Report* 26, no. 5: 34–36.

Meslin, Eric M. 1999. Engaging the public in policy development: The National Bioethics Advisory Commission Report on research involving persons with mental disorders that may affect decisionmaking capacity. *Accountability in Research* 7: 227–239.

Meslin, Eric M. 2003. When policy analysis is carried out in public: Some lessons for bioethics from NBAC's experience. In *The Nature and Prospect of Bioethics*, ed. J. Humber and R. Almeder. Humana.

Meslin, Eric M., and Summer Johnson. 2008. National bioethics commissions and research ethics. In *The Oxford Textbook of Clinical Research Ethics*, ed. E. Emanuel et al. Oxford University Press.

Meslin, Eric M. 2002. In brief: Some clues about the President's Council on Bioethics. *Hastings Center Report* 32, no. 1: 8.

Meslin, Eric M. 2004. The President's Council: Fair and balanced? *Hastings Center Report* 34, no. 2: 6–8.

Meslin, Eric M., and Harold T. Shapiro. 2002. Bioethics inside the beltway: Some initial reflections on NBAC. *Kennedy Institute of Ethics Journal* 12, no. 1: 95–102.

National Bioethics Advisory Commission. 2001. *Ethical and policy issues in international research: Clinical trials in developing countries*, volume 1. http://bioethics.georgetown.edu.

National Bioethics Advisory Commission. 1997. *Cloning Human Beings*, volume I. http://bioethics.georgetown.edu.

National Bioethics Advisory Commission. 1998. *Research Involving Persons with Mental Disorders That May Affect Decisionmaking Capacity*, volume 1. http://bioethics.georgetown.edu.

National Bioethics Advisory Commission. 1999. *Ethical Issues in Human Stem Cell Research*, volume I. http://bioethics.georgetown.edu.

National Commission for the Protection of Human Subjects of Biomedical and Behavioral Research. 1975. *Research on the Fetus: Report and Recommendation*. Government Printing Office.

National Commission for the Protection of Human Subjects of Biomedical and Behavioral Research. 1976. *Research on Prisoners: Report and Recommendations*. Government Printing Office.

National Commission for the Protection of Human Subjects of Biomedical and Behavioral Research. 1977. *Research on Children: Report and Recommendations*. Government Printing Office.

President's Commission for the Study of Ethical Problems in Medicine and Biomedical and Behavioral Research. 1981. *Defining Death: A Report on the Medical, Legal and Ethical Issues in the Determination of Death*. Government Printing Office.

President's Commission for the Study of Ethical Problems in Medicine and Biomedical and Behavioral Research. 1981. *Protecting Human Subjects: First Biennial Report on the Adequacy and Uniformity of Federal Rules and Policies, and Their Implementation, for the Protection of Human Subjects of Biomedical and Behavioral Research*. Government Printing Office.

President's Commission for the Study of Ethical Problems in Medicine and Biomedical and Behavioral Research. 1982. *Compensating for Research Injuries: A Report on the Ethical and Legal Implications of Programs to Redress Injuries Caused by Biomedical and Behavioral Research*. Government Printing Office.

President's Commission for the Study of Ethical Problems in Medicine and Biomedical and Behavioral Research. 1983. *Deciding to Forgo Life-Sustaining Treatment: A Report on the Ethical, Medical, and Legal Issues in Treatment Decisions.* Government Printing Office.

President's Commission for the Study of Ethical Problems in Medicine and Biomedical and Behavioral Research. 1983. *Implementing Human Research Regulations: Second Biennial Report on the Adequacy and Uniformity of Federal Rules and Policies, and of Their Implementation for the Protection of Human Subjects.* Government Printing Office.

President's Commission for the Study of Ethical Problems in Medicine and Biomedical and Behavioral Research. 1983. *Securing Access to Health Care: A Report on the Ethical Implications of Differences in the Availability of Health Services,* volume 1. Government Printing Office.

Raab, Harriet. 1999. Memorandum to Harold Varmus, M.D., Director, NIH, Federal Funding for Research Involving Human Pluripotent Stem Cells, January 15.

Shapiro, Harold T., and Eric M. Meslin. 2005. Relating to history: The influence of the National Commission and its *Belmont Report* on the National Bioethics Advisory Commission. In *Belmont Revisited,* ed. J. Childress et al. Georgetown University Press.

Spielman, Bethany. 2003. Should consensus be "the commission method" in the US? The perspective of the Federal Advisory Committee Act, regulations and case law. *Bioethics* 17, no. 4: 341–356.

Tuskegee Syphilis Study Ad Hoc Advisory Panel. 1973. Final report. Public Health Service. http://biotech.law.lsu.edu.

U.S. Congress, Office of Technology Assessment. 1993. *Biomedical Ethics in U.S. Public Policy—Background Paper.* Government Printing Office.

World Health Organization, Department of Ethics, Trade, Human Rights and Health Law. Interactive bioethics commission map. http://www.who.int.

IV

Conflicting Views of Biotechnology

9
Technoprogressive Biopolitics and Human Enhancement

James J. Hughes

Progressive politics is rooted in the history of the European Enlightenment. It combines a faith in the possibility of human progress with a commitment to values of individual freedom, social equality, solidarity, democratic governance, and the supremacy of reason over dogma and tradition. Progressive politics is rooted in the idea of individual citizens governing their own lives and, where necessary, enacting accountable government through discussion to pursue the common good.

Reaction to the advance of the Enlightenment project has occurred on many fronts, from clerics objecting to challenges to their authority to elites defending their privilege. Reaction has also come from within the Enlightenment, from those who believed one aspect or the other of the Enlightenment, such as individual liberty, the growth of democratic state, or the triumph of reason, threatened other, more important values. The present-day biopolitics of enhancement medicine, itself a direct result of the Enlightenment, is one of the many forms in which this struggle over the Enlightenment project and progressive political values has manifested. The project of progressive bioethics is to pursue Enlightenment values in health care and biopolicy, and to defend them against both external and internal critics.

These guiding Enlightenment principles are clear in the four core values of the Center for American Progress' Bioethics project (2006):

Human dignity: Promote the ability of individuals to achieve a sense of their unique worth and pursue their vision of the good life.

Critical optimism: Support a science that improves our lives, frees our imagination, and is responsive to our human values.

Equity: Ensure equal access to the benefits of modern society, including health care and medical technology.

Social justice: Support social and economic policies that respect and protect the lives and health of all people.

In other words, liberty and solidarity ("human dignity"), reason ("critical optimism"), and equality ("equity" and "social justice").

Jonathan Haidt's (2007) work on the evolutionary psychological origins of moral and political values provides another frame in which to understand the conflict of Enlightenment values with their critics. Haidt argues that progressives (he uses the word 'liberals') are principally motivated by two basic moral intuitions: the desire for justice and the desire to help others and prevent harm. Conservatives, however, are motivated by three additional moral intuitions: respect for authority, in-group loyalty, and ideas of spiritual pollution and purity. Rejecting the relevance of these latter three values, progressives assert instead the importance of self-determination, universalism, and reason. If Haidt is correct, while progressives are not free from innate moral intuitions, they have focused on several core values consistent with Enlightenment thought, to the exclusion of moral intuitions consistent with pre-Enlightenment ethics.

Another way to understand the topology of values is to observe how they shape political movements and parties. Empirically, political movements in the industrialized world in the twentieth century were defined along two broad axes: economic politics and cultural politics. Economic conservatives are generally opposed to the social welfare state, trade unions, taxation, business regulation, and economic redistribution. Economic progressives generally favor these measures as means to ensure fairness and the public good (i.e., moral impulses toward justice and helping others). Cultural conservatives generally are nationalists, ethnic chauvinists, religious conservatives, and oppose women's equality, sexual freedom, and civil liberties (i.e., moral impulses toward in-group solidarity, authority, and spiritual purity are valued over individual liberty and universalism). Cultural progressives are more secular and cosmopolitan, and support civil liberties and minority rights. This allows us to parse movements and parties into one corner or another of a two-dimensional terrain, or somewhere in between (figure 9.1).

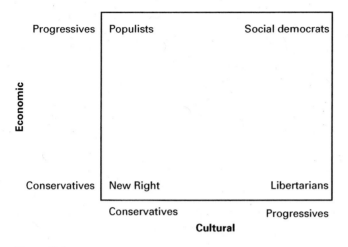

Figure 9.1
The political axes of twentieth-century politics.

The emergence of biotechnological controversies gives rise to a new axis, biopolitics, not entirely orthogonal to the previous dimensions but certainly distinct and independent of them. Allies along one or two dimensions may find themselves opposed on biopolitical issues (figure 9.2).

Biopolitics is defined by advocacy for or rejection of the biotechnologies such as enhancement medicine, with "transhumanists" (Bostrom 2005)—those who believe that humans should be able to make consensual use of biotechnology to pursue life, happiness, ability, and progress—at the progressive end of the spectrum, and "bioconservatives" at the other. Right bioconservatives are generally motivated by pre-Enlightenment values of religious authority, in-group solidarity, and fear of spiritual pollution, while left bioconservatives are (at least putatively) motivated by concerns about equality, liberty, and the public good.

The polarization between the transhumanist and bioconservative positions within biopolitics is manifest in a variety of different contexts:

Who is a citizen with a right to life?   Who is owed the rights of citizenship, such as the right to life, is central to biopolitical debates over abortion, stem cells, great apes, brain death, and the creation of human-animal chimeras. While biopolitical progressives assert that all intelligent

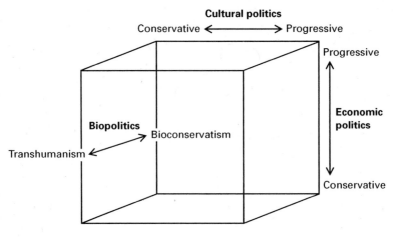

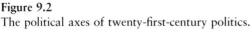

Figure 9.2
The political axes of twenty-first-century politics.

?? "persons" are deserving of rights, whether human or not, bioconservatives insist that "humanness" is the basis of citizenship and rights.

Control of reproduction Bioconservatives are critical of efforts to
control reproduction, from religious objections to contraception, abortion, and fertility treatments to unease over genetic testing, germ-line
gene therapies, and cloning. Biopolitical progressives defend the use of
reproductive technologies on grounds of bodily autonomy and reproductive rights.

Fixing disabilities and "human enhancement" Bioconservatives are
anxious that efforts to reduce the incidence of disability in society—e.g.,
through prenatal screening, prosthetics, and therapies—will have bad
consequences for children, the disabled, and society. They also are
anxious that technologies that enhance abilities above the norm will
degrade social equality and violate sacred values. Biopolitical progressives defend efforts to reduce the incidence of disability and efforts to
enable consensual use of enhancement technologies.

Extending life Bioconservatives generally defend a "natural" limit
to human longevity and reject radically extended life through anti-
aging drugs and therapies. Biopolitical progressives defend radical
prolongevity.

Control of the brain   Bioconservatives decry the effects neurotechnologies may have on virtue, equality, and autonomy. Biopolitical progressives generally defend the right of individuals to use neurotechnologies, such as psychopharmaceuticals and brain chips, to achieve greater happiness and ability.

As these biopolitical debates become increasingly relevant to daily life, they are being engaged by increasingly broad segments of the public, and biopolitical alliances and ideologies are forming around them.

Using this framework for biopolitics, I want to outline a history of the ideas in the upper front right corner of figure 9.3, at the nexus of the progressive politics of culture, economics, and biopolitics—the "technoprogressive" point of view (A). The other principal group on the transhumanist side is the libertarian transhumanists (B). I will also discuss the two principal groupings of bioconservatives, those who come from the political left (C) and those from the cultural and economic right (D).

The relatively new term 'technoprogressive' has been growing in use among progressives who also support the consensual use of safe

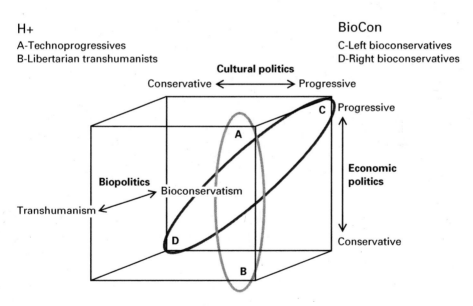

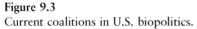

Figure 9.3
Current coalitions in U.S. biopolitics.

enhancement technologies, especially those associated with the Techno-liberation list and website. One prominent exponent of the term 'techno-progressive' is Dale Carrico, a lecturer at the University of California at Berkeley. In Carrico's (2006) formulation, technoprogressives "assume that technoscientific developments should be and can be democratizing, sustainable, and emancipatory so long as they are regulated by legitimate democratic and accountable authorities to ensure that their costs, risks and benefits are all fairly shared by the actual stakeholders to those developments. Technoprogressive stances variously support such techno-scientific development in general, and tend to take up strong positions of support for informed, nonduressed consensual human practices of genetic, prosthetic, and cognitive modification in particular."

While Carrico emphatically does not consider himself a transhumanist, many others who use the term do, and in that usage 'technoprogressive' overlaps the "democratic transhumanism" I articulated in my 2004 book *Citizen Cyborg*. As in previous progressive debates about the bourgeois deviationism and the subcultural irrelevance of communitarianism, sexual libertinism, and countercultural affectation, technoprogressives are divided about identification with the transhumanist movement. Differences also relate to the biopolitical model above, as purists like Carrico reject the idea of a relatively autonomous biopolitical axis and see no rationale for technoprogressives to tactically ally with libertarians or left bioconservatives. For those who identify as left transhumanists, on the other hand, there are strategic possibilities for technoprogressives to ally with both libertarians and left bioconservatives around specific issues. In both Carrico's usage and my own, however, technoprogressivism is the consistent application of progressive values to technology and enhance-ment issues. Survey research shows that progressives outnumber libertar-ians among self-identified transhumanists 2 to 1 (Hughes 2005), which supports the contention that support for human enhancement is as con-sistent with the egalitarian strain of the Enlightenment tradition as with its purely liberal strain.

Using these biopolitical categories, we can see the challenges facing a progressive bioethics coalition-building project. Although libertarian, technoprogressive, and left bioconservatives all agree on such Culture War issues as abortion rights, they are divided on biopolitical issues and economic issues. (See figure 9.4).

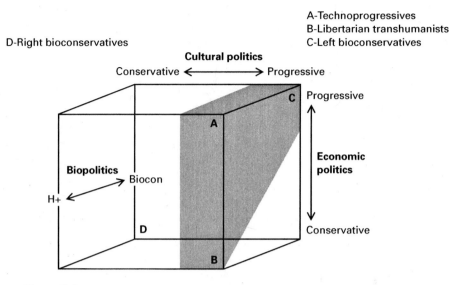

Figure 9.4
The challenge of a progressive bioethics coalition.

The Technoprogressive Thread and the New Biopolitics

From the seventeenth century through World War II, liberals and radicals were generally convinced that the progress of science and reason would free humanity from the pains and limitations that conservatives believed natural and desirable. This enthusiasm led them to champion scientific medicine, public health, and eventually universal health care. It also led them to anticipate the radical enhancement of the human body and brain, and the complete defeat of disease and death.

Francis Bacon's *Novum Organum* (1620), one of the first Enlightenment manifestoes, advocated "effecting all things possible" using science to improve the living condition of human beings. In Bacon's *New Atlantis*, medicine is used not only to eliminate disease but also to increase strength, relieve pain, retard aging, and prolong life. This enthusiasm for progressive rational mastery of the body can also be found throughout the writings of the Enlightenment *philosophes*.

In 1665, Robert Hooke proposed artificial organs, implants to enhance sight and hearing, and machines to enhance memory: "By the addition of such artificial Instruments and methods, there may be, in some manner, a reparation made for the mischiefs, and imperfection, mankind has drawn upon itself."

Diderot claimed that future science would be able reanimate the dead, take a man's brain apart and put it back together, and create human-animal chimeras and intelligent machines, and that we might evolve into posthuman forms (Hughes 2007). Leibniz wrote "In the course of time the human race may reach a greater perfection than we can imagine at present." (Bury 1980) Voltaire speculated about extending the human lifespan through medical science. Writing to Joseph Priestley in 1780, Benjamin Franklin predicted that "in a thousand years . . . all diseases may by sure means be prevented or cured, not excepting even that of old age, and our lives lengthened at pleasure even beyond the antediluvian standard."

In 1793, the physician, scientist, and Enlightenment philosopher Thomas Beddoes wrote of the eventual conquest of tuberculosis: ". . . however remote medicine may at present be from such perfection . . . the same power will be acquired over living, as is at present exercised over some inanimate bodies, and that not only the cure and prevention of diseases, but the art of protracting the fairest season of life and rendering health more vigorous will one day half realize half the dream of Alchemy." Also in 1793, the British anarchist philosopher William Godwin wrote in his *Enquiry Concerning Political Justice* that mankind would not only throw off governments and churches in the future, but that there would also be no disease, ill health, or aging. Godwin asked: "If the power of intellect can be established over all other matter, are we not inevitably led to ask, why not over the matter of our own bodies? . . . In a word, why may not man be one day immortal?"

In the aftermath of the French Revolution, Condorcet (1795) imagined that reasons would eventually defeat not only slavery, tyranny, and patriarchy but also death and the need to labor: "Nature has set no term to the perfection of human faculties; the perfectibility of man is truly indefinite; and the progress of this perfectibility, from now onwards independent of any power that might wish to halt it, has no other limit than the duration of the globe upon which nature has cast us."

In the nineteenth century and the early twentieth, many socialists believed that radical social reform would also liberate science and the body. In a 1923 essay titled "Daedalus or Science and the Future," the Marxist scientist J. B. S. Haldane predicted the adoption of human genetic enhancement and "ectogenesis" (artificial wombs) and argued that medical progress is always a challenge to orthodoxies and traditions of their day. For Haldane, biotechnologists were revolutionary Prometheans.

One of the earliest proposals for a brain prosthesis was broached by the Marxist J. D. Bernal in his 1929 essay "The World, the Flesh & the Devil: An Enquiry into the Future of the Three Enemies of the Rational Soul." For Bernal, evolution into a cyborgian posthuman diversity was a natural corollary of radical social progress: "[T]his final state would be so fluid and so liable to improve, and . . . there would be no reason whatever why all people should transform in the same way . . . to predict even the shapes that men would adopt if they would make of themselves a harmony of form and sensation must be beyond imagination. . . ."

Eugenics, the Bomb, and the Ascendance of the Luddite Left

Enthusiasm for public health and bio-utopian possibility also led some Progressives and socialists to endorse eugenic ideas and coercive reproductive policies, such as involuntary sterilization, with disastrous consequences. After World War II, reaction against the horrors of fascism tarred any consideration of a genetic approach to public health for progressives and became the first of many factors to make progressives more skeptical about technology.

In the 1960s, as progressives mobilized against nuclear weapons, the military-industrial complex, ecological destruction, and consumer culture, the romantic, pastoralist reaction against modernity became increasingly influential. The New Left inveighed against "the machine" and "technological rationality," and the counterculture attacked positivism and lauded pre-industrial ways of life. Deconstructionists and post-modernists cast doubt on the Enlightenment's "master narratives" of political and scientific progress. Deep ecologists challenged the basic humanist presumption of Enlightenment thought—that humanity gives meaning and purpose to the world—and called for a return to ecological natural law. Critics of the iatrogenic effects of physician patriarchy, the medical-industrial complex, and "Western medicine" cast doubt on the progressive promise of universal access to health care, or of any benefits to be gained from biotechnologies. After the 1960s, suspicion became the default progressive reaction to new biotechnologies, especially any technology having to do with genetics.

It was in this context that bioethics first emerged. Bioethicists questioned the use and direction of biotechnologies, establishing the rights of subjects in medical research, the rights of patients to refuse medical interventions, and the dangers of *in vitro* fertilization, cloning, and genetic engineering. Influenced by the anti-technology orientation of their generally progressive milieu, the bioethicists generally saw their role as critics of the technoscientific enterprise. However, the technoprogressive Enlightenment tradition began to re-emerge among bioethicists as early as the 1970s—for example, in the work of the Episcopalian theologian Joseph Fletcher (1974), who argued that humans have a right and an obligation to control their own genetics. For some Christian bioconservatives, the polarization between Joseph Fletcher and his fellow cleric-bioethicist Paul Ramsey over death with dignity, abortion, cloning, and genetic engineering marks the beginning of modern biopolitics.

Mass biopolitics in the United States clearly began with the *Roe v. Wade* decision, which mobilized a large army of Christian activists around a bioethical debate over the personhood of the embryo and the fetus. The abortion issue combined with animal rights and brain death in bioethics circles to crystallize points of view on the relevance of personhood and humanness to rights-bearing, an issue that was at the core of the emerging biopolitics.

Jeremy Rifkin was another explicit harbinger of the supersession of bioethics by biopolitics. In the late 1970s, Rifkin, a former socialist activist, formed the Foundation on Economic Trends to oppose bio-capitalism and any efforts at "the improvement of existing organisms and the design of wholly new ones with the intent of perfecting their performance." Rifkin quickly discovered the importance of alliances with the religious right, built on their shared critiques of biotech hubris. Rifkin built alliances between Catholic conservatives and bioconservative feminists in his campaign against surrogate motherhood, and between anti-biotech Greens and the Christian right around opposition to recombinant genetic engineering and DNA patenting. In 2001, Rifkin convinced prominent progressives to join with conservatives to call for a ban on cloning of embryonic stem cells.

Rifkin is quite clear about the importance of his odd coalitions to the coming "fusion biopolitics." In a 2001 article titled "Odd Coupling of Political Bedfellows Takes Shape in the New Biotech Era," Rifkin says: "The Biotech Era will bring with it a different constellation of political visions and social forces, just as the Industrial Age did. The current debate over embryo and stem cell research already is loosening the old political allegiances and categories. It is just the beginning of the new politics of biology."

The Kass Era

Despite these stirrings of left-right biopolitical coalitions, it was not until the appointment of Leon Kass as the chair of the President's Council on Bioethics (PCB) that the new biopolitics finally began to gel. Kass had opposed every intervention into human reproduction, from *in vitro* fertilization to reproductive cloning, and his appointment was a concession by the George W. Bush administration to the religious right, as he was certain to lead the council to condemn research on embryonic stem cells. Kass stacked the PCB with conservatives. After dutifully recommending that the use of embryos in research should be criminalized, he focused the PCB on opposition to human enhancement, from psychopharmaceuticals to life extension. That opposition resulted in the PCB's 2003 report *Beyond Therapy*.

The rise of Kass to a position of prominence was a shock to the liberal bioethics community, and their resentment was deepened by the growing realization that conservatives were pouring millions of dollars into the training of Christian right bioethicists and the formation of bioethics institutions. The Women's Bioethics Project (Hinsch 2005, 4) documented the rapid growth of Christian right bioethics organizations, and concluded as follows:

• Conservatives have well-established bioethics centers with strong advocacy outreach programs that are interlocking and supportive of each other.

• Conservatives are using an existing infrastructure of think tank and religious organizations to drive awareness, energize their constituencies, and support a unified bioethics agenda.

• Conservative foundations are strategically funding high-profile cases with a broad bioethics agenda in mind.

• Conservatives see driving bioethical debate as critical to building a society based on their values and worldview.

Since 2001, opposition to human enhancement technologies has been a central motivating cause for these conservative biopolicy groups. The first two conferences (2003 and 2004) of the Christian Right Center for Bioethics and Culture (CBC) were on the theme "TechnoSapiens: The Face of the Future." The CBC also coordinated the "Manifesto on Biotechnology and Human Dignity," which opposed abortion, cloning, and human enhancement and which was endorsed by most of the leaders of the American right. In the Midwest, the base for Christian right bioethics is Chicago's Center for Bioethics and Human Dignity (CBHD) at Trinity International University. The CBHD warned the Christian right about the threat from enhancement through conferences and through its network of hundreds of affiliated scholars and graduates. In 2006, for instance, the CBHD led a campaign involving the Christian Right Center for Bioethics and Culture, James Dobson's organization Focus on the Family, the Concerned Women of America, and the Christian Medical and Dental Association in an attack on a grant by the National Institutes of Health to the progressive bioethicist Maxwell Mehlman for work on ethical guidelines for federally funded research on genetic enhancement. The campaign ludicrously insisted that the grant was federal support for "Nazi eugenics" and "transhumanism," and that Mehlman was a prominent transhumanist.

The 2002 publication of Francis Fukuyama's critique of prospects for human enhancement, *Our Posthuman Future*, opened the door for serious consideration of human enhancement by the conservative policy establishment in Washington. One prominent base for conservative bioethics in the Washington area is the Ethics and Public Policy Center, closely tied to the Kass council and the source of the journal *The New Atlantis*, which has published influential attacks on artificial intelligence, nanotechnology, biotechnology, reproductive technology, and life extension.

Progressive critics of biotechnology also stepped up their opposition to enhancement technologies after 2001. For instance, in 2001 George Annas and Lori Andrews co-authored a piece arguing for an international treaty to make cloning and germ-line genetic therapy a "crime against humanity," a call that was taken up by the left bioconservative Center for Genetics and Society. In 2003, the liberal writer Bill McKibben weighed in with an anti-enhancement tract titled *Enough*. The progressive Canadian ETC Group began campaigning for a moratorium on nanotechnology and nano-enhancement. In 2003, Lori Andrews joined with Nigel Cameron to unite a dozen prominent progressives and a dozen religious right activists in the first "fusion" bioconservative organization, the Chicago-based Institute for Biotechnology and Human Future.

The rise of Kass, the religious right, and anti-enhancement sentiment forced some progressive bioethicists to clearly declare themselves on the side of human biological self-determination. In 2001, the bioethicist Gregory Pence published a book titled *Who's Afraid of Human Cloning?* And in 2002, Gregory Stock published *Redesigning Humans*, a defense of genetic enhancement and germ-line therapy. The progressive bioethicists Allen Buchanan, Dan Brock, Norman Daniels, and Daniel Wikler co-authored *From Chance to Choice: Genetics and Justice*, defending the possibility, even the necessity, of a liberal egalitarian approach to human genetic enhancement.

The growing influence of the right's bioethicists created the need for a distinctly progressive bioethics voice, and leading liberal bioethicists (among them Arthur Caplan, Glenn McGee, Alta Charo, Laurie Zoloth, and Jonathan Moreno) have sought to mobilize progressive bioethicists. As Kathryn Hinsch noted in her report (2005, 4), however, explicitly progressive bioethics groups have far fewer funds, fewer troops, and a lot of theoretical work to do to mount a coherent response to the right:

• What progressive activities there are in the area of bioethics are underfunded, narrowly focused, and lacking in a unified philosophical framework.

• Although progressives dominate academic bioethics, the scholars are not trained and in many cases are disinclined to work from an explicit ideological framework.

• Progressives will need to do more than throw money at the problem; it will require a major rethinking of the issues.

Among these problems, the "rethinking of the issues" is probably the most pressing. Although agreeing on the need for some form of universal health care and reproductive rights, cultural and economic progressives disagree about many of the emerging biopolitical issues. Do reproductive rights include prenatal screening and genetic engineering? Are people with disabilities liberated by efforts to cure and ameliorate their disability, or do those efforts only oppress them further? Is the prospect of human enhancement a fulfillment of the progressive vision of human self-emancipation, or the road to a caste society?

Table 9.1 lists biopolitical organizations addressing human enhancement.

Three Meta-Policy Contexts for Technoprogressive Approaches to Enhancement

There are three principal meta-policy contexts that will shape policy toward human enhancement technologies in the coming decade: the ongoing erosion of health insurance in the United States and the need for a guarantee to basic universal health care, the demographic pressures of the aging of industrial societies, and globalization.

The U.S. Health-Care Crisis

Universal health care has been a part of the U.S. progressive agenda since the Progressive movement a hundred years ago. It is clear that another wave of support for fundamental reform is building, helped a Democrat win the presidency in 2008, and will likely be a central domestic agenda item for Congress. Since the last major effort to reform our privately managed health-care system, in 1994, we have been unable to contain inflation in the cost of health care, and the steady increase in the per capita and proportional costs of health care is hurting the sectors of the

Table 9.1
Biopolitical organizations addressing human enhancement.

	Important personalities	URL	Description
Religious bioconservatives			
Center for Bioethics and Culture Network	Jennifer Lahl	thecbc.org	Loose Christian right network
Center for Bioethics and Human Dignity	John Kilner	cbhd.org	Runs influential training program and conferences at TIU
Ethics and Public Policy Center	Eric Cohen Adam Keiper	eppc.org	Beltway religious conservatives, tied to Kass; publish *The New Atlantis*
Center for Nanotechnology and Society	Nigel Cameron	nano-and-society.org	Opponent of nano-enhancement
Discovery Institute	Wesley Smith	discovery.org	Vocal opponent of evolution and human enhancement
Libertarian transhumanists			
Reason magazine	Ron Bailey	reason.com	Leading libertarian journal; Bailey is prominent defender of enhancement technologies
Foresight Institute	Chris Peterson	foresight.org	Advocate for molecular manufacturing and nano-enhancement
Left bioconservatives			
Center for Genetics and Society	Marcy Darnovsky Richard Hayes	genetics-and-society.org	Left opponents of "techno-eugenics"
ETC Group		etcgroup.org	Opposed to genetic engineering and nanotechnology for safety and equity reasons

Table 9.1
(continued)

	Important personalities	URL	Description
Not Dead Yet	Stephen Drake	notdeadyet.org	Radical disability group; opposed to enhancement technologies
Friends of the Earth	Brent Blackwelder	foe.org	Active in building bioconservative coalition against enhancement
Technoprogressives			
Center for Responsible Nanotechnology	Mike Treder Chris Phoenix	crnano.org	Advocates for regulated, safe, egalitarian nano-enhancement
Center for Cognitive Liberty and Ethics	Wrye Sententia Richard Boire	cognitiveliberty .org	Advocates for "cognitive liberty"
Institute for Ethics and Emerging Technologies	J. Hughes	ieet.org	Technoprogressive virtual think tank
IHEU Appignani Center for Humanist Bioethics	Ana Lita	iheu.org	Lobbyist for humanist bioethics at UN
Alliance for Aging Research	Daniel Perry	agingresearch .org	Lobbying for research into anti-aging therapies
Left-right fusion bioconservatives			
Institute for Biotechnology and a Human Future	Lori Andrews Nigel Cameron	thehumanfuture .org	Largely defunct; superseded by CNS
Left-libertarian fusion transhumanists			
World Transhumanist Association	Nick Bostrom	transhumanism .org	Leading transhumanist group; chapters worldwide

U.S. economy that have foreign competitors with much lower health-care costs.

A single-payer system is a still a long shot, and the political rationales for a regulated universal health-care voucher system such as that proposed by the Clinton reform effort of 1994 are still in place. Whether progressives are given the opportunity to rally behind a single-payer or a universal-voucher proposal, there will be a need to define what the basic level of coverage The Plan or all of the offered plans must provide. The benefits to be included in the basic level of care were a contentious issue for the Clinton plan in 1994, with vigorous debate about mental health benefits for instance. Inclusion of emerging enhancement technologies will be a part of any future reform debate as well, dividing technoprogressives and left bioconservatives over the relevance of the therapy/enhancement distinction, an argument that has raged over issues including cosmetic procedures, sex reassignment, attention-deficit disorder, and fertility treatments. A technoprogressive approach to priority setting would ignore dubious therapy/enhancement distinctions and instead rely on cost-benefit analyses such as the Quality Adjusted Life Year. Therapies that provide a high per-dollar return in QALYs (which takes into account life-expectancy and the quality of remaining life-years) would be included in a basic guaranteed level of coverage, while those providing fewer QALYs per dollar would be out-of-pocket.

Demographic Shifts and the Longevity Dividend
Industrialized societies are all facing some degree of structural adjustment as the number of retired seniors proportional to working taxpayers increases. In the United States this has been a poisonous issue for legislators, as the Bush administration and the Republican Congress discovered when floating the idea of Social Security privatization in 2004–05. Progressives probably will continue to rally behind Social Security, and probably will attack the idea that the system is facing a crisis. The prospect of enhancement technologies will play a critical role in the debate, however, as age-retarding therapies increasingly offer the prospect of extending the period of healthy, disability-free longevity—a "Longevity Dividend" (Olshansky et al. 2006). The incidence of disability and chronic illness among seniors in the United States is declining steadily,

leading the Census Bureau to extend its estimates of when Medicare and Social Security will be insolvent. Age-retarding therapies offer the possibility of reducing age-related illness and disability further, reducing the need for drugs and nursing care, and keeping seniors and their family care givers in the labor market.

Anti-aging therapies are already a widely popular idea, and the retiring Baby Boomers are sure to support a Longevity Dividend program of federal investments in anti-aging research, and the provision of effective therapies through Medicare or its successor. Although a technoprogressive advocacy for anti-aging therapies will be popular, this will again be an issue of contention for left bioconservatives, who are likely to question the feasibility and desirability of such therapies.

Globalization

When Italy enacted draconian controls on assisted reproduction in 2004, restricting the right of lesbians and single women to fertility treatments for instance, Italian women with sufficient resources simply sought treatment in more liberal parts of Europe. Transgender people in the United States increasingly travel to countries such as Thailand for sex re-assignment surgery, for one-fourth the price in the United States and without the need to wait through a year of psychiatric evaluation. Brazil, Bolivia, Mexico, Colombia, Costa Rica, Argentina, Israel, Singapore, Malaysia, India, the Philippines, and South Africa have become magnets for Americans seeking inexpensive cosmetic and elective surgeries, for as little as one-tenth the cost in the United States. Some U.S. employers, facing mounting health insurance premiums, have encouraged employees to seek medical treatments abroad.

The globalization of medical treatment poses a major challenge to efforts to regulate human enhancement technologies, especially for progressives. If prenatal genetic and anti-aging treatments, or novel prosthetics and body modifications, are restricted in the United States, there are sure to be less-regulated providers somewhere else in the world. To the extent that restrictions on enhancement technologies are imposed in one country, they will penalize only the poor, not the affluent (who can afford to travel abroad).

Globalization also will influence the debate over human enhancement in the area of economic competitiveness. The rise of the Indian and

Chinese economies, combined with their investments in higher education and emerging technologies, their large populations, and their relatively low wages, ensure that they will increasingly draw investment capital away from the United States and compete against U.S. goods in international trade. In the 1990s, Robert Reich argued that the U.S. could compete in the global economy by upskilling U.S. workers to take on increasingly high-tech and knowledge-intensive work. Unfortunately, the rising cost of American higher education, and meager federal aid, has caused the United States to fall behind many other industrialized countries in the production of graduates, especially in math, science, and engineering.

As enhancement therapies become increasingly efficacious, they will also influence economic competitiveness, directly by extending the abilities and productivity of workers and indirectly by adding workers to the labor force who would otherwise have been disabled. India and China, lacking any notable bioconservative resistance to enhancement technologies, will especially welcome the prospect of using enhancement technologies to facilitate their economic growth. The pressure of global competition will thus also likely encourage liberal, universal access to enhancement technologies in the United States.

Who Are Potential Technoprogressive Constituencies?

The New Deal coalition has eroded since the 1970s, torn apart by the Democratic Party's unwillingness to mobilize the middle class and the poor with populist economic appeals and the New Right's success in appealing to the cultural conservatism of the religious, the poor, and white men. The Congressional Progressive Caucus has been a consistent and growing voice for a social democratic re-orientation of the Democratic Party, one which has much promise in winning back groups alienated by the Culture War. However, if my analysis of emergent biopolitics is correct, rebuilding a majority coalition for progressive reform will require not only a re-emphasis on economic populism but also a conscious strategy of biopolitical appeals. Just as segments of the public were alienated from the Democrats when they were caricatured as effete, disdainful, atheist, gay abortionists, segments of the religious right are already testing the mobilizing effect of labeling the left as promoters of the Brave New World, while libertarians deride liberals and the Food and Drug Administration for standing in the way of life-saving medicine.

Articulating a technoprogressive approach to biopolitics offers both a secure philosophical basis and a consistent policy basis for making popular biopolitical appeals to important constituencies, helping to build their support for a broader progressive movement.

Seniors

Seniors are the first and most obvious constituency to whom a guarantee of universal access to safe enhancement medicine, particularly anti-aging medicine, will appeal. Progressives have a natural case to make that the bioconservatives' insistence that seniors get sick and die on time is profoundly ageist. On the other hand, the prospect of successful prolongevity also poses challenges for progressives, as we will need to argue for a progressive re-negotiation of the work life, pensions, and the retirement age. But the alternative is a breakdown in generational equity and solidarity, and potentially of the safety net itself.

The Disabled

People with disabilities, using the latest assistive technologies and with their eyes fixed on medical progress, are another natural constituency for technoprogressive advocacy for enhancing technologies. By appealing to the vast majority of disabled who strongly support enabling cures and prosthetics, progressives can marginalize the few, but vocal, radical disability activists who reject enhancing technologies as neo-eugenic. This is especially true since these radicals aligned with the right during the Terri Schiavo controversy. Although issues such as prenatal screening and cochlear implants will likely remain difficult, strong commitments to research into, and access to, curative and enabling technologies will likely overcome these qualms.

Supporters of Reproductive Rights

Bioconservative feminists have found it difficult to convince women that some choices they might make about the contents of their wombs are not included in reproductive rights, and for good reason. Although activism on behalf of contraception and abortion has rarely included demands for freedom of germinal choice and access to artificial reproductive technologies, these are two sides of the same coin. The struggle for reproductive rights has been technoprogressive from its outset, a struggle for universal access to safe enabling technologies that permit control of human reproduction in novel, "unnatural" ways. Today the idea that only parents, and not the state, should make reproductive decisions is broadly popular. Even support for parents' rights to prenatal genetic enhancement and sex selection is growing, as documented by surveys by the Center for Genetics and Public Policy. By embracing a full conception of reproductive rights, including the right to ensure the fullest health and ability for one's children with safe and accessible enhancing technologies, a technoprogressive approach can appeal broadly to parents.

Advocates of Drug-Law Reform

The War on Drugs has been an enormous obstacle to progressive reform, wasting public finances on the prison-industrial complex and militarizing police forces in communities of the poor and people of color while doing nothing to reduce the burdens of chemical dependency. In the coming decade, therapies to treat and prevent chemical dependencies will, it is to be hoped, shift the debate in favor of a public-health approach to illegal narcotics. But progressives cannot wait for these therapies, and must embrace drug-law reform as a central, albeit less popular, issue.

Rational drug-law reform, based on sound research on harm, will also strongly influence the regulation of cognitive enhancement drugs such as Modafinil. While there is still broad support for a criminal approach to psychopharmaceuticals, distinguishing between those with low risks (such as cannabis and cognitive enhancers) and those with high risks (such as methamphetamine) is an approach that polling suggests many people will understand.

Health Lobbies and the Scientific Community

The Republican war on science (Mooney 2005), in particular the battle over research on embryonic stem cells, has pushed many patient advocacy groups and scientific lobbies in Washington toward the Democrats. These groups share a broad interest with technoprogressives in seeing more public financing of medical research and in protecting the freedom to conduct research from bioconservative bans.

Sex-Gender Nonconformists

Gays, lesbians, bisexuals, and transgender individuals are a natural technoprogressive constituency with intrinsic interests in enhancing technologies. The champions of natural law attack sex-gender nonconformity and human enhancement with the same arguments. Gays and lesbians have already been victimized by bioconservative laws, such as those in Italy, denying them access to reproductive technology. Access to safe, subsidized cosmetic, hormonal, and (potentially) genetic therapies is a central issue for transgenders.

In short, the technoprogressive appeal, defending the right to use regulated and widely accessible emerging technologies in enabling ways, is a vital component in the construction of a popular and philosophically consistent progressive coalition.

Technoprogressive Approaches to Funding, Regulating, and Providing Enhancing Technologies

Ensure universal access to all beneficial biotechnologies through universal public provision   A technoprogressive approach acknowledges that emerging and enhancing technologies can exacerbate inequality. But technoprogressives believe that the best way to ameliorate this risk is to ensure ever greater access to the benefits of enabling technologies, as with literacy, laptops, and health care. Progressives have never argued for the banning of expensive but beneficial medical therapies, such as antiretroviral therapies for HIV, but rather have sought ways to make them accessible to everyone who needs and wants them.

Eliminate the therapy/enhancement distinction in research funding and healthcare   Progressives should set priorities for research funding, regulation, and provision using utilitarian rubrics such as the Quality Adjusted Life Year (QALY), not on the basis of shaky, pre-modern ideas of normative health. A QALY-based assessment should include both "therapies" and "enhancements" in a basic guaranteed tier of insurance if the "therapies" and "enhancements" provide a high level of QALY per dollar.

Regulate enhancement technologies for safety, not morality   Techno-progressives argue for vigorous and independent regulation of emerging technologies to ensure their safety and efficacy, and other *tangible* public goods, while remaining generally neutral about the life goals that guide their consensual use. Technoprogressives reject bioconservative demands for moral regulation of biotechnology, or for the banning of technologies on the grounds that they might have vaguely defined ill effects on the family or on social solidarity.

Defend cognitive liberty with rational drug-law reform   The 2006 UK Science and Technology Select Committee review of British drug laws in relation to the latest assessments of risks associated with drugs is a model for technoprogressive reform. Under a harm-based regulatory system, access to cognitive enhancement drugs would likely be liberalized, along with access to low-risk recreational drugs such as cannabis and MDMA.

Federally fund research into the biology of aging and other enhancement targets   Many areas of basic science research contribute to innovation in enabling and life-extending technologies without being targeted at those goals. But federal programs that specifically target enhancement will speed the pace of development. One such technoprogressive initiative is the recent call by a broad coalition of biogerontologists and public policy scholars (Olshansky et al. 2006) for the National Institutes of Health to create a specific "Longevity Dividend" program. The proposal for a "Cognos" brain-science initiative made by the Nano-Bio-Info-Cogno project (Bainbridge and Roco 2002) of the National Science Foundation is another example.

Roll back intellectual-property overreach, especially patenting of the human genome   The legally groundless patenting of the human genome—which is not an invention—stands in the way of further bio-medical progress, enhancing or otherwise. Technoprogressives oppose this corporate overreach and seek a far more conservative standard in intellectual property in general to protect fair use and the sharing of information.

Establish cognitive personhood as the basis of rights-bearing under the law   Humanness is a reactionary, pre-Enlightenment standard on which to base rights-bearing. Technoprogressives advocate for a cognitive standard for personhood, relevant in abortion, embryo research, brain injury and brain death, genetic enhancement, the rights of animals and human-animal chimeras, and potentially in adjudicating the status of machine-augmented humans and machine minds.

Conclusions

As progressive bioethicists work to articulate their policies and beliefs, they find themselves divided by the emerging biopolitics. Insofar as left bioconservatives are strictly concerned with the safety of therapies and with their equitable distribution, these concerns can be addressed by a technoprogressive program of thorough and independent regulation and a universal health-care system. Insofar as bioconservative concerns are motivated by deeper suspicions about the Enlightenment project that technoprogressivism and human enhancement represent, however, a common progressive bioethics program is unlikely. A technoprogressive approach to human enhancement is merely the consistent application of the values that have been at the core of progressive political movements since the Enlightenment: the right of individuals to be free to control their own bodies, brains, and reproduction according to their own conscience, within democratic nations that work for the public good. If progressives and progressive bioethicists adopt this consistent approach, they can make popular biopolitical appeals to important constituencies, and can build a majority coalition in support of progressive change.

Works Cited

Agar, Nicholas. 2004. *Liberal Eugenics: In Defence of Human Enhancement.* Blackwell.

Annas, George J., Lori Andrews, and R. M. Isasi. 2002. Protecting the endangered human: Toward an international treaty prohibiting cloning and inheritable alterations. *American Journal of Law & Medicine* 28: 151–178.

Bacon, Francis. 1905. Novum Organum. In *The Philosophical Works of Francis Bacon,* ed. J. Robertson. Routledge.

Beddoes, Thomas. 1793. *Letter to Erasmus Darwin M.D. on a New Method of Treating Pulmonary Consumption and Some Other Diseases Hitherto Found Incurable.* Bulgin and Rosser.

Bainbridge, William, and Mihail Roco. 2002. Converging technologies for improving human performance: Nanotechnology, biotechnology, information technology, and cognitive sciences. http://wtec.org.

Bernal, J. D. 1929. *The World, the Flesh & the Devil: An Enquiry into the Future of the Three Enemies of the Rational Soul.* Dutton.

Bostrom, Nick. 2003. A transhumanist perspective on human genetic enhancements. *Journal of Value Enquiry* 37, no. 4: 493–506.

Bostrom, Nick. 2005. A history of transhumanist thought. *Journal of Evolution and Technology* 14, no. 1: 1–25.

Buchanan, Allen, Dan Brock, Norman Daniels, and Daniel Wikler. 2002. *From Chance to Choice: Genetics and Justice.* Cambridge University Press.

Bury, John B. 1980. *The Idea of Progress: An Inquiry into Its Origin and Growth.* Basic Books.

Carrico, Dale. 2006. Technoprogressivism: Beyond technophilia and technophobia. *Amor Mundi*, August 12, 2006. http://ieet.org.

Center for American Progress. Progressive Bioethics. http://www.americanprogress.org.

Condorcet, Nicolas Marquis. 1795. Sketch for a historical picture of the progress of the human mind. http://oll.libertyfund.org.

Firestone, Shulamith. 1970. *The Dialectic of Sex.* Morrow.

Fletcher, Joseph. 1974. *The Ethics of Genetic Control: Ending Reproductive Roulette.* Doubleday.

Franklin, Benjamin. 1999. Letter to Joseph Priestley. In *Theological and Miscellaneous Works of Joseph Priestly*, ed. J. Rutt. Thoemmes Continuum.

Fukuyama, Francis. 2002. *Our Posthuman Future.* Farrar, Straus and Giroux.

Godwin, William. 1793. *Enquiry Concerning Political Justice.* http://dwardmac.pitzer.edu.

Haidt, Jonathan, and Jesse Graham. 2007. When morality opposes justice: Conservatives have moral intuitions that liberals may not recognize. *Social Justice Research* 20, no. 1: 98–116.

Haldane, J. B. S. 1923. Daedalus or science and the future. http://www.cscs.umich.edu.

Hayes, Richard. 2007. Our biopolitical future: Four scenarios. *Worldwatch Magazine* 20, no. 2: 10–17.

Hinsch, Kathryn. 2005. *Bioethics and Public Policy: Conservative Dominance in the Current Landscape.* Women's Bioethics Project.

Hooke, Robert. 1665. *Micrographia, or, Some Physiological Descriptions of Minute Bodies Made by Magnifying Glasses.* J. Martyn and J. Allestry.

Hughes, James. 2004. *Citizen Cyborg.* Westview.

Hughes, James. 2005. Report on the 2005 interests and beliefs survey of the members of the World Transhumanist Association. http://transhumanism.org.

Hughes, James. 2007. Dreaming with Diderot. January 14. http://ieet.org.

McKibben, Bill. 2003. *Enough: Staying Human in an Engineered Age.* Times Books.

Mooney, Chris. 2005. *The Republican War on Science.* Basic Books.

Olshansky, S. Jay, Daniel Perry, Richard Miller, and Robert Butler. 2006. In pursuit of the longevity dividend. *The Scientist* 20, no. 3: 28–36.

Piercy, Marge. 1976. *Woman on the Edge of Time.* Fawcett.

Rifkin, Jeremy. 2001. Odd coupling of political bedfellows takes shape in the new biotech era. *Los Angeles Times,* July 24.

Science and Technology Select Committee. 2006. *Science and Technology—Fifth Report.* House of Commons.

Smith, Wesley J. 2002. The transhumanists: The next great threat to human dignity. *National Review Online,* September 20. http://www.nationalreview.com.

Stock, Gregory. 2003. *Redesigning Humans: Choosing Our Genes, Changing Our Future.* Mariner Books.

10

Biopolitics, Mythic Science, and Progressive Values

Marcy Darnovsky

When bioethical deliberation confronts human biotechnologies, it often faces novel questions that recent technical developments have conjured into existence. Many of these biotechnology-related situations are both socially consequential and, at least in some respects, unprecedented in human experience. After spending its first years considering these issues largely as a scholarly endeavor with practical applications in human research and medical practice, bioethics in recent years has ventured far beyond academic and clinical settings.

Among the bioethical issues that have lately become topical are cloning and stem cell research; regulation of the assisted-reproduction industry, "designer babies" and the prospect of a market-driven "techno-eugenics," sex selection and disability de-selection, surrogacy and reproductive tourism, patents on life, and markets in kidneys, eggs, and other human tissues. Formerly in the provinces of philosophers, physicians, and lawyers, these topics are now frequently discussed in the mainstream media, in popular culture, and in electoral politics. They are of interest to entrepreneurs, to venture capitalists, and to global industries. They have captured the attention of policy makers, political consultants, candidates, philosophers, ministers, and rabbis. They are subjects for social theorists, political pundits, talk show hosts, moviemakers, and fiction writers.

Bioethicists themselves have expanded their range. Some now work (on salary or as paid consultants) for biotechnology companies, civil-society organizations, and political campaigns. This spillover of bioethical concerns into public awareness reflects the extent to which bioethics has come to encompass large and pressing questions about the role of biological sciences and technologies in democratic society and human life. In short, bioethics has become biopolitics.

For those committed to expanding the purview of democratic governance and the scope of concerns that citizens and civil-society organizations examine, the transformation of bioethics to biopolitics is a welcome development. But the birth of biopolitics has been a difficult one. A number of biopolitical issues have fallen into the deep cultural and partisan divides of our day. Some are now used as litmus tests, with conversations about them resembling exercises in ideological positioning. This polarized environment has not been conducive to thoughtful consideration of the issues that biopolitics raises for progressives.

More generally, biopolitics has also become part of an increasingly contested politics of science. The practical and conceptual issues raised by human biotechnologies will resemble to some extent those that confront environmentalists, public health advocates, and others whose issues are enmeshed in the development and use of powerful technologies. In engaging and assessing all these endeavors, we progressives can and should draw deeply on our commitments to social justice and the common good, to public-sector oversight, to a precautionary sensibility in the face of powerful technologies, and to the broad inclusion of civil society in democratizing science and science policy.

The Politics of Science in the Bush Era

Liberals and progressives engaged in the politics of science faced unusual challenges during the presidency of George W. Bush. His administration demonstrated early and often that it would not hesitate to bend science policy to its purposes, or to use science selectively for its political and ideological advantage. The president and his appointees clashed with mainstream scientific opinion on climate change, attacked evolution, and meddled aggressively in environmental protection, drug regulation, and reproductive health. Yet there was funding aplenty for military science and defense technologies.

Thankfully, this egregious pattern focused the attention and energies of progressives and liberals. Scientists, research advocates, Democratic politicians, progressive pundits, and public-interest organizations documented and decried the Bush administration's habit of twisting science policy in the interests of its favored constituents and benefactors.

Unfortunately, however, some aspects of this much-needed challenge were off target. In their efforts to counter the Bush administration's propagandistic uses of science, some progressives wound up adopting or acceding to assumptions that could undermine rather than support a progressive politics of science. These unfortunate tendencies are of particular concern in the fraught areas of genetic, reproductive, and biomedical technologies.

"War on Science" as a Flawed Metaphor

Because the Bush administration was aggressive in selectively using science and abusing science policy, the temptation to dub its objectionable policies a "war on science" was understandable. This rhetorical move efficiently mobilized scientists and their supporters. But as is often the case with military metaphors, it hijacked careful thinking. Beyond a sound-bite horizon, it created more problems than it solved.

First, the "war on science" framework suggested that Bush and his team treated science policy differently than other policy arenas—the law, the arts, labor, international relations, education, and so on. In fact, the Bush administration was confrontational in all these fields, calculating its policies to satisfy its corporate funders and its voter base. "The Bush administration," science policy expert Daniel Greenberg pointed out, "has interests—ideological, theological and compliant to some industries—that are its preoccupations. Scientists have an inflated sense of themselves if they think the administration has anything against them in particular as it pursues its goals in ways that disregard their views." (quoted in Vergano 2007)

The "war on science" metaphor also blurred a politically crucial distinction between policies and politics motivated by the Bush administration's alignment with the Christian right and those that were driven by the agendas of its corporate allies. These disparate motives, of course, mirror the uneasy coalition between free-market policies and social conservatism that had propped up the Republican Party since the 1980s. To win elections, to fund campaigns, and to govern, Republican candidates and officials had to please both corporate elites and Christian fundamentalists. Thus the Bush administration carefully attended to the preferences of corporate America—especially Big Oil, Big Construction, Big Pharma, and military-related industries—and strenuously upheld its ideological commitments to promoting the market and minimizing public-interest regulation. These priorities explain most of Bush's objectionable science-related policies: the years of denying the findings of climate research, the Orwellian approach to environmental problems, the relentless efforts to undercut food and drug regulation, and the vengefulness toward whistleblowers.

But the Bush team also hewed to the preferences of its Christian fundamentalist base. In the science policy arena, this meant distorting research on abstinence-only education, overruling the advice of scientific panels to block over-the-counter access to emergency contraception, raising questionable doubts about the efficacy of condoms in preventing the spread of HIV/AIDS, appointing unqualified but ideologically compatible figures to important positions that involve matters such as abortion and contraception, and, of course, restricting federal funding for embryonic stem cell research. The policy on such funding, which represented one of the few instances in which the Bush administration declined to give any corporate interests pretty much what they wanted, stood in stark contrast to its pro-corporate, anti-regulation bias in many other areas of environmental and science policy.

Unfortunately, liberal and progressive opponents of the Bush stem cell policy typically accepted the Christian right framework, in which controversy over the moral status of human embryos was the primary concern about the endeavor. In part because of the "science wars" narrative, they reflexively embraced what the enemy supported, and raised no questions about conflicts of interest, commercial imperatives, exaggerated promises, and effective oversight of stem cell research. In so doing, they forfeited the opportunity to craft a pro-research stance that would also have highlighted the need for consistent and enforceable regulation and oversight, for hope without hype, and for close attention to principles of social justice.

The military metaphor also misleads in another way: it recruits us as defenders of "science" instead of suggesting that we carefully evaluate the social implications of the matter at hand. Because the Bush administration shamelessly used science as a partisan wedge, Daniel Sarewitz remarked, "the opportunity to use science as a political tool against Bush has been irresistible—but it is very dangerous for science, and for politics. You can expect to see similar accusations of the political use of science in the next regime." (quoted in Vergano 2007)

Positioning ourselves as science's white knights can too easily suggest a simplistic view that portrays contemporary scientific practice as a pure and value-free enterprise, innocent of social and economic power. The military metaphor slides too easily into a Manichean construction of policies or people as either "for" or "against" science. This is a notion for which there is little empirical referent. Do even the Amish forgo eyeglasses or yogurt? Does Dick Cheney oppose nuclear reactors or precision-guided bombs? It is far less helpful, conceptually or politically, to decry the Bush administration as having been "anti-science" than it is to examine the way that partisan, ideological, and corporate agendas can make selective use of scientific findings and distort science policy. It is far less meaningful to ask whether one is for science than to ask what kind of science one is for.

At times, progressive and liberal defenders of science seemed to suggest that developments in science and technology more or less exhaust the meaning of human progress, that they always redound to the benefit of humanity and never to its shame or sorrow, and that science exists outside of human choices and human efforts, independent of social, cultural, and economic dynamics. From these starting points, it would follow that scientific inquiry is—or should be—isolated from social and political values, above the fray of political and economic power. In this view, the only ethical issues germane to an assessment of science would be those that pertain to conduct (or misconduct) inside laboratories and professional journals. Other ethical questions, along with social issues of all sorts, would be considered only in evaluating the technologies that are the eventual products of scientific investigation. These assumptions are, at best, an ideal-type way of understanding science. As a description of contemporary scientific endeavor, they are better described as mythic.

The Problems with Mythic Science

The mythic view of science that has taken hold among some progressives can be understood in large part as a reaction to the Bush administration. Some who promote this view are probably aware of its limitations; a bit of probing might well reveal far more complex and nuanced ideas about the relationship between science and society. But some proponents of mythic science seem ready to defend it as a matter of principle. That spirit suffuses a number of books that excoriate religion in the name of science, including those by Richard Dawkins, Christopher Hitchens, Sam Harris, and Daniel Dennett. More than one reviewer has noted that those books have an "evangelical" flavor (Johnson 2006; Prothero 2007). In the *New York Times*, George Johnson (2006) described a November 2006 forum on religion and science that "began to resemble the founding convention for a political party built on a single plank: in a world dangerously charged with ideology, science needs to take on an evangelical role, vying with religion as teller of the greatest story ever told."

One need not partake in the arguments that pit science against religion to question a mythic view of science. Nor is it necessary to accept a particularly strong version of what scholars call "the social construction of science" to acknowledge the deep and inextricable ways in which contemporary science is shaped by social and cultural values, economic interests, and power dynamics. This is not just a matter of philosophy or epistemology. Historical and sociological studies, as well as journalistic investigations, show that there are no longer sharp distinctions between science as a disinterested confrontation with nature and technology as the application of the knowledge thus acquired to practical matters. This intensifying entanglement of science and technology over the past century is captured by the term 'techno-science'. As Dan Sarewitz put it (2006), "the distinction between science and technology is increasingly blurred, to the point where it confuses more than it clarifies."

The mythic view of science is also belied by the movements of resistance that have managed to shape the practice and products of techno-science. Environmentalists, who have mounted one of the most effective and popular social movements of the past 50 years, rely heavily on scientific findings and also work to influence the direction of scientific inquiry. They fight against some technological applications and promote others. Few would argue for unfettering science from social and environmental values; the basic insights of environmentalism counsel care in the selection of technologies and attention to their social and ecological impacts.

Challenging the mythic view of science is in no way at odds with appreciating that scientific investigation is a powerful way to produce shared and reliable information about the world, or with reveling in the wonders of scientific methods and scientific knowledge, or with enthusiastic support for many of the products of scientific endeavor and technological development. Nor does it in any sense condone government, partisan, or corporate suppression or alteration of research findings or scientists' statements.

The importance of rejecting the mythic view lies in recognizing that techno-science is a social endeavor that is inseparable from social dynamics and enterprises. The particular knowledge produced by even the most basic research is a result of decisions made about what sorts of knowledge to pursue. These broad decisions are made by people and institutions—researchers and research communities; universities, companies, and non-profit research institutes; philanthropic, government, and venture funders. Neither the agendas, the methods, nor the products of science stand above the social world. The practitioners and funders of scientific research have interests, affiliations, and values. The outputs of scientific research transform the way many people are born and live, work, and die. They create winners and losers, and enable some people to make decisions that will shape the lives and life chances of others, perhaps on the other side of the world or in future generations.

Techno-science has created delightful gadgets and wondrous tools; it has also caused environmental degradation, displaced millions of people from their homes and livelihoods, and enabled genocide. It seems implausible to deny that many decisions about techno-science—about the allocation of resources, about accountability and transparency in the conduct of research and the development of technologies, about regulation and oversight—are appropriately political issues. This makes it unsettling when progressives couch their criticisms of the Bush administration's selective uses of science as denunciations of "politicizing science." Whether this rhetoric is deployed for short-term political convenience or adopted as part of a mistaken mythic notion of science as truth, it is conceptually flawed and politically inadequate. It is likely to lead to wrongheaded politics and dangerous policies.

What happens, for example, when we progressives want to bring our values to decisions about science and science policy? Shouldn't social justice, human rights, and the public interest inform our approach to scientific matters? Shouldn't "responsible science" connote active commitments to evaluating the social meanings and implications of particular research directions, and to incorporating democratically shaped visions of human progress into research programs and policies?

The challenge for democratic societies—and it is not an easy one to meet—is to protect science from illegitimate partisan interference while working to make science, especially publicly funded science, accountable to progressive values and to a democratic polity. We need to eschew the inappropriate *politicization* of science. At least as urgently, we need to find ways to *democratize* it.

Progressive Values and the Politics of Science

It is not difficult to imagine biotechnological developments and social contexts that would help produce a world in which we have less and less commitment to one another as members of a single human community, in which the divide between the haves and the have-nots increasingly and perhaps irreversibly deepens. Who will profit, who will lose, and who will survive in the biotech century? Celebrity scientists? Biotech entrepreneurs? Attractive college students whose eggs are in demand? Athletes—at the professional, amateur, and high school levels—tempted or pressured by "gene doping?" People with Down Syndrome? Villagers in South Asia who sell their kidneys or rent their wombs to wealthy North Americans? And who will decide? What rules will there be, and who will enforce them? Progressive approaches to these questions—and to many others posed by high-tech reproduction and biomedicine—will require careful thought about several broader political questions:

1. How can we ensure that social justice, the public interest, and the common good are given place of privilege in the development of human biotechnologies, while protecting individual autonomy?

2. How can we enact responsible societal regulation of genetic, reproductive, and biomedical technologies, rather than leave their development and dissemination to laissez-faire market mechanisms and corporate agendas?

3. What kinds and degrees of enthusiasm and caution will progressive biopolitics bring to its assessments of various biotechnologies?

4. How can progressive biopolitics encourage democratic deliberation about and civil-society involvement in decisions about powerful new biotechnologies?

These are not exclusively biopolitical questions; they crop up in other political arenas as well. But they are particularly pertinent to twenty-first-century biopolitics in the United States, where powerful economic and ideological forces push in the direction of a libertarian sensibility, a market orientation, and an unnuanced enthusiasm about techno-science.

The United States has long seen strong bipartisan support for science and technology as engines of economic development on the road to societal betterment. In 1945, when Vannevar Bush defined science as "the endless frontier," he signaled a commitment not just to the pursuit of scientific knowledge, but also to a narrow definition of progress as scientific and technological advance, and to a prioritization of individual and market-based solutions to social problems. As we enter a new technological era in which the biological rather than the physical sciences are promoted as the location of limitless horizons and exciting economic opportunities, it is time to re-examine those assumptions.

Social Justice, the Public Interest, and the Common Good

Most progressives have a strong commitment to social justice, solidarity, and the common good on one hand, and on the other place high value on individual autonomy, liberty, and freedom. These commitments must often be balanced; in some cases, priorities must be determined and choices made. Bioethicists have explicitly addressed the need to navigate this tension; two of the often-cited four basic principles of bioethics developed by Tom Beauchamp and James Childress (1979) are "autonomy" and "justice." But in negotiating the always tenuous balance between these principles, bioethics has often tilted in the former direction.

Especially when bioethics focuses on the ethical quandaries posed by high-tech medical treatments in clinical settings, it tends to favor expanding the autonomy of individual patients in narrowly defined environments and creating procedural rules such as those to ensure the informed consent of patients and participants in clinical trials. Also pulling bioethics toward an emphasis on individual autonomy is the increasingly potent consumer model of medical care, especially strong in societies that lack a publicly funded health system. The field of assisted reproduction, for example, which has posed many controversial questions for bioethics, exists in the United States predominantly in the private sector. Most patients pay the high costs of fertility treatment out of pocket, and with price considerations necessarily looming large a consumer mentality is hard to avoid.

The bioethics tradition has always also included concern about the larger social and policy implications of new biological technologies. Some bioethicists, especially many with a feminist sensibility, have worked to make social justice and human-rights concerns more central to the field's theoretical, conceptual, and practical concerns. But a stronger emphasis on individual autonomy has been common, and this inflection must be taken into account in the shift from bioethics to biopolitics that is now underway.

Another skew toward the individual-rights end of the autonomy-justice spectrum is found in the decades-long controversy in the United States over abortion, and in the way that abortion rights have been framed and defended. Unfortunately, we have not yet succeeded in establishing abortion as a fundamental right that is based both on our desire to promote the dignity and well-being of women and on our determination that women's freedom to terminate unwanted pregnancies will contribute to the kind of society we want to build. Instead we have framed abortion rights in public understanding as an issue of "choice" (which can too easily sound like a matter of consumer preference) and in legal precedent as a matter of privacy (too fragile a judicial reed).

These frameworks—themselves in part results of ongoing attacks on the legality of abortion and access to it—have made it difficult to see or to concede that some questions about reproduction call for a logic that moves beyond privacy and choice. A woman's decision to terminate an unwanted pregnancy is very different from the development, marketing, and use of technologies that can alter the biological traits of future children and future generations. Sam Berger put it as follows in "A Challenge to Progressives on Choice" (2007):

Biotechnological innovations . . . are quickly shifting certain reproductive decisions from matters of private choice to ones of public concern, regardless of the moral status of fetuses and embryos. Parents in the twenty-first century will have the ability to control the genetic makeup of their children in ways that were unthinkable fifty years ago. The choices they make will thus significantly affect the structure of society. As progressives, we must acknowledge the new challenges posed by these reproductive technologies and, when necessary, craft policies to limit their potentially harmful impact.

In these and other biopolitical matters, the progressive values of social justice, the public interest, and the common good provide important guidance to our assessment of the new biological technologies.

New Divides and Discrimination

The first-order questions about social justice concern equity and distributive justice, the rich and the poor. Can a particular biotech product or practice be made broadly accessible? Will its application preferentially increase the privilege of the already wealthy, or will it aid the most vulnerable among us? Bringing these concerns to bear doesn't mean that those who can afford treatments and procedures should automatically have to forgo them, or that as a society we should eschew all biomedical advances until everyone in the world has caught up. But it does compel our attention to the equity dimensions of biopolitics. It does oblige us to look skeptically at developments that amount to "designer medicine" (available to the few, unaffordable by the majority). It should lead us to heed Dorothy Roberts's call to "imagine a multi-billion-dollar industry designed to create Black children" (1997).

Similarly, decisions deciding how to allocate research funds and how to encourage research talent should not be approached as a zero-sum situation. But shouldn't we prioritize clean drinking water for the third of the world's population that doesn't have it, and ensure health care for the 46 million uninsured Americans? In the fight against malaria, doesn't it make sense to provide mosquito nets before spending huge sums of money on genetically modified mosquitoes or on anti-malarial agents produced through synthetic biology?

As progressives committed to the fight for universal health care, we will in any case soon have to face difficult decisions about the kinds of high-cost medicine that such a program can support. Our commitment to the common good will at times persuade us to put public health and broad accessibility ahead of costly biotechnologies.

Consider again the policy questions raised by assisted reproduction. Infertility is a devastating condition for many people, and millions have successfully overcome it with *in vitro* fertilization and related techniques. But in the United States, where fertility treatment is inadequately covered even for those with health insurance, poor people typically can't afford it—despite the fact that the obstacles they face in getting primary medical care means that they are less likely to be treated for conditions that can cause infertility. During the 1980s and the 1990s, when people with means were increasingly using reproductive technologies, poor and immigrant women were being castigated for having too many children. As Roberts points out (1997), the parents of sextuplets born after fertility treatments became feel-good media celebrities; mothers of poor children were dubbed "welfare queens."

Universal health coverage that includes access to basic medical care and to fertility treatment could actually mitigate some of the medical expense, since prevention is less costly than treatment. It will also help because patients with coverage for fertility treatment will be less desperate to succeed on the first IVF attempt, and therefore less likely to request or accept transfers of many embryos, which lead to the high-order multiples whose care costs millions of dollars. Nonetheless, more publicly funded fertility treatment will increase the need to set limits.

Ensuring fair access to reproductive or other biotechnologies is one dimension of social justice. Sometimes, social justice instead entails setting limits. Some uses of human biotechnology in fact exacerbate existing discrimination and disparities, or create new forms of exploitation and inequality.

Already people from wealthy countries are traveling to India for surrogacy arrangements. Brokers recruit poor women from rural villages who will carry a pregnancy for a fraction of the cost of a woman closer at hand, and will be far less likely to change their minds about relinquishing the baby. In a global economy, we can expect even more new twists on exploitation. And surrogacy is also a class issue at home. By 2008, it had become clear that the economic crisis was causing a surge in the number of young women seeking payment for providing eggs or a womb. Brokers in the assisted-reproduction business in the United States acknowledge that they look for surrogates in less wealthy areas of the country, and describe the most important characteristic of a candidate as "compliancy" (Mundy 2007). It is hardly in keeping with progressive values to support a new manifestation of class division, and one based on old forms of gender discrimination to boot.

Genetic screening of embryos and fetuses poses additional concerns, expressed most forcefully by disability-rights advocates who see themselves as targets of selection technologies. They are acutely aware that procedures enabling the selection of "good" genes and "normal" traits can devalue the bodies—and ultimately the lives—of people with disabilities. Newer genetic and reproductive technologies under development—and already widely used in laboratory and farm animals—raise the prospect of market-driven, high-tech eugenics, in which assisted reproduction coupled with genetic modification of early embryos would produce "enhanced children" for the elite. Scattered voices are already openly promoting this "designer-baby" scenario.

Most proponents of developing designer-baby technology (technically known as "inheritable genetic modification") acknowledge that its expense would confine it to the most privileged, and that it would become a powerful new force of inequality and exclusion. One book co-authored by several bioethicists and health-policy experts considered a range of scenarios that would make inheritable enhancements widely available, or would prevent them from generating unprecedented inequality. The best available option, the authors concluded, would be a lottery that would give poor people a chance at qualifying for genetically enhanced offspring. One advantage of this plan, they noted, would be ensuring that the lower classes maintained their belief in the possibility of upward mobility, important as a stabilizing force in capitalist democracies (Buchanan et al. 2000). It is troubling that the prospect of inheritable genetic modification can tempt this kind of misinterpretation of the meaning of equality and democracy.

Human biotechnologies may also affect racial justice in unwanted and unintended ways. Certain biotech applications, or the emphasis they place on technological rather than social or environmental factors, could disproportionately burden minority communities. More subtly, and at least as consequentially, some biotechnology studies and products are being promoted in ways that could revive discredited ideas about race. Dorothy Roberts (2004) writes:

Perhaps the greatest danger posed by the biotech agenda is its power to intensify racial injustice in America. Not only are human biotechnologies being employed within a racist social order, but they are already reinforcing the myth that race is a genetic trait and impeding efforts to tackle the social causes of racial inequality.

Remember, justifying racial inequities in biological terms rather than in terms of white political privilege has profoundly shaped science in America for three centuries, beginning with the scientific defense of slavery. This basic explanation of racial difference rooted in biology rather than power still operates today—in fact it is making a spectacular come back—and threatens to shape every aspect of the biotechnological future.

In a study titled "Playing the Gene Card?: A Report on Race and Human Biotechnology," Osagie Obasogie, a Senior Fellow at the Center for Genetics and Society (my organization), discusses the impacts on minority communities of three biotechnology-based products: race-specific pharmaceuticals (such as BiDil, approved as a treatment for heart failure in African Americans by the FDA in 2005), DNA forensic databases, and genetic ancestry tests. Obasogie emphasizes the important difference between the use of race as a biological category—often in ways that attribute racial inequality to genetic variations—and the use of race as a political tool in order to remedy disparities that stem from social and economic causes. He cautions that "given our unfortunate history of linking biological understandings of racial difference to notions of racial superiority and inferiority, it would be unwise to ignore the possibility that 21st century technologies may be used to revive long discredited 19th century theories of race" (Obasogie 2009).

Democratic Governance and Corporate Science

The need for active oversight and regulation in the public interest by mechanisms of democratic governance is a central theme in the history of progressive ideas. As concentrations of capital and corporate power grew in the late nineteenth century and in the twentieth century (in part because of contemporaneous technical developments), progressives' commitment to an affirmative role for government strengthened.

The twenty-first century has been called "the biotech age," and observers expect significant growth driven by the biological sciences and technologies. These new biotechnologies are taking shape in an environment that is commercialized and market-driven to an unprecedented extent, and in the hands of researchers with previously unheard-of levels of direct interest in profit making and corporate gain.

A generation ago, scientists were far less likely than they are today to be involved with private industry as consultants, stakeholders, and founders. The norms of academic science, which in the past provided a fairly high degree of openness, objectivity, and commitment to a broadly defined public interest, are being undermined by the increasing involvement of university-based researchers in private companies.

Many widely acknowledged problems are easily traceable to the increasingly blurred line between academic and corporate research, and to the turn away from effective government regulation and oversight. Conflicts of interest are rife among biological researchers; expansive awards of intellectual property are being made, a free-market intellectual property regime is being imposed on the rest of the world, adverse reactions in clinical trials go unreported because of proprietary concerns, and the pharmaceutical and biotechnology industries exert increasing economic and political clout over both U.S. political parties. More and more studies document significant proportions of researchers who have withheld data or delayed publication of their work because of personal financial concerns. We have fewer and fewer researchers whose loyalty to their work and to scientific advance is untouched by financial investments and interests. These commercial ties are increasingly accepted, even lionized—scientist-entrepreneurs have become celebrities.

Of course, many scientists remain ethical, responsible, and devoted to expanding knowledge and developing tools to benefit humanity. But we urgently need a new ethos and new rules to expose and minimize conflicts of interest, to restore a critical mass of scientists free of financial ties, to disentangle academic research from corporate influence, and to strengthen public-interest science.

Fifty years ago, Jonas Salk became famous for developing the polio vaccine. In April 1955, just after the announcement that the field trials of the vaccine had been successful, and in the midst of national celebrations that greeted the news, Salk was interviewed by Edward R. Murrow on the television program *See It Now*. Murrow asked: "Who owns the patent on this vaccine?" Salk replied: "Well, the people, I would say. There is no patent. Could you patent the sun?" (Smith 1990) Salk's response is almost unimaginable today. We live in an age of corporate biotechnology, and we urgently need to develop effective ways to protect the public interest.

It is neither possible nor desirable to subject basic scientific research to a litmus test of whose interests the results will serve. But with the distinction between basic science and its practical applications increasingly blurred, progressives should be cultivating an appreciation for the imperatives of profit and power. The emergence of techno-science and the ever-closer relationships between university and commercial science calls for close attention to the substitution of corporate agendas for public-interest objectives. Yet it is increasingly common for scientists working with new biological technologies, and even for some progressives involved in the politics of science, to argue that voluntary self-regulation is sufficient. Professional, trade, and *ad hoc* organizations now routinely issue "guidelines" for research that are unenforceable and sometimes widely flouted. In a recent issue of the *Bulletin of the Atomic Scientists*, for example, Drew Endy (a pioneer of synthetic biology) acknowledged that engineering techniques applied to microbial genomes pose the risk of releases (deliberate or inadvertent) of virulent pathogens against which human populations have no immunity. For Endy, even this frightening prospect does not constitute a persuasive case for increased oversight and regulation. Endy's prescription is to distribute knowledge about synthetic biology techniques as widely as possible, and hope that in the case of a bioterror attack someone will quickly cook up an antidote (Endy 2007).

With regard to reproductive and genetic technologies, it is widely acknowledged that the United States has gaping holes in regulation and oversight. There is no comprehensive policy of the sort that has been established in many other countries. (Those of Canada and the United Kingdom are discussed below.) The Food and Drug Administration and the Centers for Disease Control exert some oversight in specific areas, and a patchwork of rules and guidelines are applied, sometimes sporadically, to certain technologies and procedures. But many decisions about new technologies of assisted reproduction—and indeed about the human genetic future—are currently being made by small groups of scientists and private companies accountable only to themselves.

Those who oppose societal regulation and oversight have been emboldened by dwindling public confidence in governmental capacity and competence. This situation must be understood as a triumph of the Republican and corporate agendas, enacted through a set of efforts and strategies that the Center for Progressive Reform has dubbed "regulatory underkill." But a majority continues to support government regulation, and that sentiment grew significantly stronger after the financial collapse of 2008 demonstrated the consequences of regulatory inadequacy.

In the commercialized conditions in which the new biological technologies are being developed, effective and responsible regulatory structures and policies are especially critical, and progressives have a vital role to play in establishing them. As in other policy arenas, we will need to protect regulatory structures from capture by corporate or narrow ideological interests. And because regulation in the biotech area will inevitably touch on questions of reproduction, we will have to be particularly alert to ensure that it will not be used illegitimately to restrict individual rights, especially those associated with abortion.

Existing Models of Comprehensive Biotechnology Governance

Fortunately, there are already models on which to build. A number of countries have established oversight and regulation of biotechnology and bioscience. The most comprehensive regulatory schemes in place are in the United Kingdom and in Canada, both countries with strong democratic traditions and clear protections for reproductive rights.

The United Kingdom's Human Fertilization and Embryology Authority (HFEA), established in 1991, licenses and monitors all research involving human embryos and all facilities offering *in vitro* fertilization or storage of eggs, sperm, or embryos. The U.K. has legislatively prohibited certain applications, including reproductive human cloning, but issues licenses for some controversial techniques, such as somatic cell nuclear transfer (SCNT, also known as research cloning).

The HFEA's 21 members are appointed by Health Ministers; at least half of them are required to be neither doctors nor scientists involved in human embryo research or infertility treatment. The agency conducts extensive public consultations on new or controversial technologies and proposed policy changes. To grant a license for research under its regulatory purview, the HFEA must be satisfied that the use of human embryos is "necessary or desirable" for a purpose enumerated in the act. The HFEA inspects licensed fertility clinics annually, produces a code of practice that guides clinics on proper conduct, and keeps a formal registry for donors, treatments, and children born. The U.K. regulatory model is permissive with regard to what it allows researchers and fertility practitioners to do, while setting and enforcing thorough rules for their conduct.

Canada's Assisted Human Reproduction Act (AHRA), approved in 2004, established stronger guidelines for permissible applications. It prohibits the creation of human embryos solely for research (including via SCNT), inheritable genetic modification and reproductive cloning, sex selection (except to prevent the birth of children with certain sex-linked conditions), and commercial surrogacy and gamete retrieval. In contrast to the United Kingdom's HFEA, the Canadian law was motivated more explicitly by commitments to the well-being and health of women and children and to preventing the commercialization of reproduction.

Several factors allowed Canadians to agree on this legislative package. Abortion rights and access to abortion services are not hotly contested in Canada; as a consequence, policies on embryo research, cloning, and related topics can be evaluated more easily on their own merits rather than with regard to their bearing on abortion politics. A network of Canadian feminists was instrumental in crafting and supporting the AHRA through the 15 years that it took to develop and pass it. Influential anti-abortion groups opposed the AHRA, though others usually associated with anti-choice positions, including the Canadian Conference of Catholic Bishops, explicitly declined to oppose it.

The Canadian Institutes of Health Research (CIHR) and other Canadian government bodies provide public funding for human embryo research, which seems to make it easier for biomedical researchers to accept public oversight and control. Many proponents of embryonic stem cell research supported the AHRA, recognizing their interest in working within a predictable and publicly acceptable framework. Another factor in the AHRA's adoption is simply the greater sense of social solidarity among Canadians, reflected in the country's universal health-care coverage and tradition of consensus politics. Also important is the fact that Canadians have been discussing the need for assisted-reproduction policies since the early 1990s, when the Canadian Royal Commission on New Reproductive Technologies launched an ambitious series of public meetings and hearings that involved thousands of Canadians from all walks of life.

A recent proposal for establishing a U.S. regulatory structure to oversee human biotechnologies, based in part on the U.K. and Canadian experiences, has been put forward by Francis Fukuyama and Franco Furger in a 400-page report titled *Beyond Bioethics* (Fukuyama and Furger 2006). Though Fukuyama associated with leading neoconservatives in the 1990s, he broke with them over the war in Iraq, and he has repudiated the neoconservative movement. In a review of *Beyond Bioethics*, Richard Hayes notes that Fukuyama, in his 2002 book *Our Posthuman Future*,

endorsed using biotechnology to address medical needs but argued against its use to modify the human species in ways that would undermine the common human nature that supports human values, behaviors, and institutions [and] irrevocably and increasingly deepen the divide between the world's haves and have-nots. His line of argument was thus closer to that of center-left critics of unrestrained genetic technology (such as Daniel Callahan of the Hastings Center, Lori Andrews of Chicago Kent School of Law, and environmentalist author Bill McKibben), than it was to that of most religious conservatives, who ground their positions largely on a belief in the personhood of the human. (Hayes 2007)

Fukuyama and Furger acknowledged the novelty—and the difficulty—of proposing a new U.S. regulatory agency during the Bush administration, particularly in the fraught area of human biotechnology. But even then, having considered the alternatives—maintaining the status quo, accepting proposals for self-regulation by researchers and the biotech and assisted-reproduction industries, passing inflexible federal laws on a technology-by-technology basis—they concluded that a comprehensive approach is the only viable method.

Techno-Exuberance versus the Case for Precaution

The new biological technologies require the application of one of the most important lessons of the environmental movement: the imperative to treat with caution powerful new technologies that may cause significant harm. The "precautionary principle" is meant to address situations of scientific uncertainty in which there is potential for significant risk—clearly an apt description for a number of new and emerging biotechnologies. The complexity and unpredictability of biological systems underlies the need for a precautionary approach.

Some criticize the precautionary principle—and even a precautionary sensibility—as a back door to opposing new technologies in general. This is inaccurate; the precautionary principle is a way to give public health and the environment higher standing than short-term commercial interests in the face of decisions about technological developments and choices among technological alternatives. Included in its purview are considerations about the common good of present and future generations—concerns that are typically lacking from traditional cost-benefit analyses that focus on economic criteria. One of its central assumptions is that parties who impose risks on others or on the biosphere should not be given primacy in policy debates. The precautionary principle is championed by environmentalists and other progressives who are cognizant of the ways that new technologies can exacerbate concentrations of power, harm the less powerful, and damage or even destroy ecological systems that support us all.

A divergent impulse in the progressive tradition emphasizes instead the liberating aspects of technology—its potential to expand human knowledge, advance standards of living, cure diseases, and minimize drudgery. Either tendency, of course, can be taken to an extreme. Among U.S. progressives today, the mythic view of science that has developed in reaction to the Christian right and the Bush administration threatens to spill over into a naive and uncritical techno-utopianism. Campaigns by self-described transhumanists for the "enhancement" of the human species through genetic and other technologies provide one example.

The techno-utopian temptation is evident in a 2007 *New York Review of Books* article by Freeman Dyson that promotes a complete remaking of the natural world via genetic manipulation and synthetic biology. Dyson looks forward to the imminent day when these biological tools become as common as iPods and cell phones. They will be used, he imagines, to produce plants with black silicon leaves that absorb extra sunlight and that provide energy too cheap to meter. Genetically engineered earthworms will imbibe their silicon detritus while extracting gold from seawater, thus putting an end to rural poverty. Dyson acknowledges that such biotech applications would pose "real and serious" dangers, but says explicitly that those dangers, along with questions about how to regulate the tools that would confer such world-changing powers, need not be considered: "I leave it to our children and grandchildren to supply the answers."

There is clearly a great expanse between such excesses and a reasonable optimism about shaping new technologies in the public interest. That is the ground on which progressives can base their biopolitical values and find wide political support.

Environmentalists and other progressives who support a precautionary approach readily support beneficial applications of new technologies, and acknowledge that we already enjoy many of them. But we must also be mindful of many reasons for concern, from the great civilizations that have collapsed because of misuse of contemporaneous technologies, to the twentieth-century genocides that depended on harnessing the scientific and medical establishments of that era, to the severe environmental degradation that plagues many parts of the world, to the prospect of globally catastrophic climate change. In a 2004 paper titled "The Precautionary Principle: A New Legal Standard for a Technological Age," Roberto Andorno puts it as follows: "Far from being antithetical to science or to technological innovation, the precautionary principle aims at promoting alternative modes of development—'safer and cleaner technologies'—in order to ensure a good quality of life for present and future generations." Andorno recounts the rapid development of the precautionary principle as a principle of national and international law, beginning with explicit references to it in a 1971 German statute on environmental protection. From there it spread to the legal systems of Denmark, Sweden, France, and other European countries in policy on food safety and public health, as well as environmental matters. Since its inclusion in the 1992 Maastricht Treaty, Andorno says, "it has become one of the pillars of the [European Union's] environmental law." In 2000 the European Commission issued a communication on the precautionary principle to provide guidelines on its meaning and appropriate applications (Commission of the European Communities 2000). At a global level, the precautionary principle has been "included in virtually every recently adopted international treaty and policy document related to the protection of the environment," including, perhaps most famously, the 1992 Rio Declaration on Environment and Development and the 2000 Cartagena Protocol on Biosafety. These national laws and international treaties contain slightly varying formulations of the precautionary principle. All these statements, however, support what Christopher Schroeder, writing for the Center for Progressive Reform, refers to as the authority of democratic governments to "intervene with respect to risky actions while there is still uncertainty about whether those actions will cause harm." And, Schroeder writes, while there are important controversies even among environmentalists about how the precautionary principle should be applied, the "main battlegrounds . . . have industry and business interests on one side and advocates of better environmental, health and safety protection on the other."

Civil-Society Governance in the Biopolitical Age

In the pre-biopolitical era, bioethicists often functioned as surrogates for democratic participation in decisions about biotechnologies. Though bioethicists typically present themselves as experts rather than spokespeople, there was and continues to be a tendency to believe that they are representing the interests of the rest of us. Particular bioethicists may or may not be doing a good job of this, but in any case bioethicists have no warrant to take it on alone. Inadvertently, their role may perpetuate a civil-society deficit on issues of human biotechnology.

Of course, democracies need experts. But bioethical expertise is a peculiar kind of expertise. Consider the role of experts in the politics of other powerful technologies. Think, for example, about controversies over what kind of energy technologies we should develop. Should we emphasize nuclear power or solar power? Should we build large hydroelectric dams? Should we drill for oil in the Arctic? On these issues, we consult (or should consult) people who are knowledgeable about the technical, policy, and social aspects and implications of the proposed technology or application. We don't consult energy ethicists.

We recognize these questions as political issues, as decisions that will transform the way many people live and work, that will create winners and losers, and that involve some people making decisions that will shape the lives and life chances of others, though they may live on the other side of the world or in future generations. We expect that these issues will be widely debated; that environmental, consumer protection, and human-rights groups are likely to weigh in, sometimes quite vociferously; that these civil-society actors will be cited in the media; and that they will be included in policy debates and decisions. We need similar robust debate, democratic participation, and broad inclusiveness of civil-society constituencies in decisions about the directions and governance of the new biological technologies. Fortunately, we are starting to see that.

Reproductive-rights and women's-health leaders, with leadership from advocates for reproductive justice, increasingly promote the view that controversy about abortion is but one aspect of reproductive politics and a broader biopolitics. Prominent feminists and national reproductive-choice organizations are taking part in discussions that start with the recognition that unequivocal support for abortion rights is consistent with regulating the biotechnology and assisted-reproduction industries in the public interest. Disability-rights advocates, early champions of precaution in the development of genetic technologies and of limits in the pursuit of genetic selection, are weighing in on these matters. Advocates of racial justice are pointing to the need to be wary of genetic discrimination and of the potential resurgence of biological explanations for racial disparities. Environmentalists are considering the long-range consequences of human biotechnologies in order to minimize risks to the public and the rest of the natural world. International health, development, human rights, and indigenous-rights organizations are challenging the push by the global biotechnology industry to place human genomics at the center of the global health agenda. Leaders of human-rights organizations and progressive religious organizations are opposing a free-market eugenics that could drive unprecedented racial, ethnic, and class division.

Bringing civil-society voices into biopolitics is a crucial endeavor, but a challenging one. Our democratic processes are imperfect. The mechanisms by which civil-society constituencies assert themselves are often messy. We have suffered through years of a politically polarized environment that makes thoughtful deliberation about human biotechnologies difficult. The combined power of techno-science and the market is daunting. But these dynamics make broad democratic inclusion and organized citizen advocacy all the more important. The efforts of civil-society organizations are sorely needed if we are to reap the benefits and avoid the perils of the new biological sciences.

Democratizing Science: A Path for Progressives

To get ourselves on the road to a progressive biopolitics, we need to affirm the need for effective and responsible policies, for inclusive democratic governance and civil-society participation in the development and deployment of biotechnologies, and for grounding our policy positions in social justice, the common good, and the public interest. The highly polarized situation in the United States makes this road look at least bumpy if not downright treacherous. The major constituencies who have until recently dominated political activity on these issues are sharply opposed, with Christian conservatives who want bans on all embryo research in one corner and biotechnology researchers and companies who want a free hand in the other. But surveys suggest that a majority of Americans prefers a middle position, one that supports human biotechnologies but wants them to be developed and used responsibly, with appropriate levels of social control and accountability. Most Americans want cures for diseases but are wary of profit-driven drug companies and unaccountable researchers. They support assisted reproduction but are increasingly aware that it is a profit-making business operating with inadequate oversight, putting women and children at unnecessary risk. In poll after poll, they oppose technologies that would alter the genetic future of the human species—that is, reproductive cloning and inheritable genetic modification—by overwhelming majorities. Both in the service of politics and in the service of principle—to appeal to public sentiment and to honor our core values—progressives should take the lead in promoting and establishing effective regulation and oversight of human biotechnologies, a strong commitment to social justice and the common good in assessing biotech applications, and a precautionary approach that reflects both well-founded public wariness and hopes for a sustainable future. Building on existing policy models, we can develop a biopolitical agenda that will win majority support and protect a progressive human future.

Works Cited

Andorno, Roberto. 2004. The precautionary principle: A new legal standard for a technological age. *Journal of International Biotechnology Law* 1, no. 1: 11–19.

Assisted Human Reproduction Canada. About the agency. http://www.hc-sc.gc.ca.

Beauchamp, Tom, and James Childress. 1979. *Principles of Biomedical Ethics.* Oxford University Press.

Berger, Sam. 2007. A challenge to progressives on choice. *The Nation*, July 18.

Buchanan, Allen, Dan W. Brock, Norman Daniels, and Daniel Wikler. 2000. *From Chance to Choice: Genetics and Justice.* Cambridge University Press.

Bush, Vannevar. 1945. *Science: The Endless Frontier.* Government Printing Office.

Center for Genetics and Society. Canada: The Assisted Human Reproduction Act. http://geneticsandsociety.org

Center for Genetics and Society. Summary of public opinion polls. http://geneticsandsociety.org.

Center for Progressive Reform. 2005. Regulatory underkill. http://www.progressivereform.org.

Commission of the European Communities. 2000. Communication from the Commission on the precautionary principle, February 2. http://ec.europa.eu.

Dyson, Freeman. 2007. Our biotech future. *New York Review of Books*, July 19.

Endy, Drew. 2007. Interview. *Bulletin of the Atomic Scientists*, May–June. http://thebulletin.metapress.com.

Fukuyama, Francis, and Franco Furger. 2006. *Beyond Bioethics: A Proposal for Modernizing the Regulation of Human Biotechnologies*. Johns Hopkins University Press.

Hayes, Richard. 2007. A majoritarian proposal for governing human biotechnology. *Bioethics Forum*, January 3. http://www.bioethicsforum.org.

HFEA website. http://www.hfea.gov.uk.

Horgan, John. 2005. Political science. *New York Times*, December 18.

Johnson, George. 2006. A free-for-all on science and religion. *New York Times*, November 21.

Mooney, Chris. 2005. *The Republican War on Science*. Basic Books.

Mundy, Liza. 2007. *Everything Conceivable*. Knopf.

Obasogie, Osagie K. 2009. *Playing the Gene Card? A Report on Race and Human Biotechnology*. Center for Genetics and Society. http://www.geneticsandsociety.org.

Prothero, Stephen. 2007. The unbeliever. *Washington Post*, May 6.

Roberts, Dorothy. 1997. *Killing the Black Body: Race, Reproduction, and the Meaning of Liberty*. Pantheon.

Roberts, Dorothy. 2004. Race and the biotech agenda. Paper presented at symposium of Center for Genetics and Society titled The Next Four Years, the Biotech Agenda, the Human Future: What Direction for Liberals and Progressives? http://genetics.live.radicaldesigns.org.

Sandel, Michael. 1996. *Democracy's Discontent*. Harvard University Press.

Sandel, Michael. 1982. *Liberalism and the Limits of Justice*. Cambridge University Press.

Sarewitz, Dan. 2006. Public science and social responsibilities. *Development* 49: 68–72.

Schroeder, Christopher. The precautionary principle. http://www.progressiveregulation.org.

Smith, Jane S. 1990. *Patenting the Sun: Polio and the Salk Vaccine*. Morrow.

Spar, Debora. 2006. *The Baby Business: How Money, Science, and Politics Drive the Commerce of Conception*. Harvard Business School Press.

Union of Concerned Scientists. Alphabetical List of Case Studies from the A to Z Guide to Political Interference in Science. http://www.ucsusa.org.

Union of Concerned Scientists. Science Idol. http://www.ucsusa.org.

Vergano, Dan. 2007. Science vs. politics gets down and dirty. *USA Today*, August 5.

Virginia Commonwealth University. Annual Life Sciences Surveys. http://www.vcu.edu.

V

Progress beyond Politics?

11

Can Bioethics Transcend Ideology? (And Should It?)

Arthur L. Caplan

For at least two decades the field of bioethics has been at the center of the ideologically driven culture wars. As the right-to-life movement gained in organization and power in the Reagan, Bush One, and Bush Two administrations, both at the federal and state levels, many issues central to the bioethical canon—abortion, end-of-life care, assisted suicide, contraception, access to infertility services, disability rights, the care of newborns, organ donation, and the use of enhancement technologies—became defining issues for both the right and the left. Discontent with secular bioethics has led to the creation of think tanks and university programs with explicit ties to evangelical Christianity or neoconservative institutions. The battle over the fate of Terri Schiavo exemplified the political importance of bioethics with the dispute involving the highest levels of the federal government and court (Caplan et al. 2007). Arguably the loss in standing of conservative views on domestic values issues can be traced to the public's discontent with the actions of conservatives and neoconservatives in intervening in the case.

Many bemoan the fact that bioethics is now both a topic of political struggle and heavily politicized (Callahan 2003). The older image of a bioethics that stood neutral as a field while individual scholars took particular points of view is yielding to a world of politicized think tanks, commission, advisory panels, councils, and foundations. Although bioethics has not become completely political, and although the old model of a "pure" apolitical bioethics is not dead, it is fair to say that the fights of the past two decades are likely to ensure that the older model of purity will not be returning anytime soon (Caplan 2008).

What roles ideology and religion ought to play in determining the policies and practices of biomedicine in the world's most powerful state is not easy to answer. But that this fight is occurring is indicative of something else, something many in bioethics and outside the field are loathe to admit: bioethics is now a mature field, and is in a position of power in American society.

Both sides in the current bioethics culture war deny that they are to blame for any excesses when it comes to disagreements. They will usually claim abiding respect for the other side while maneuvering as best they can behind the scenes with a well-placed phone call, a strategically timed op-ed piece, or a dismissive blog comment.

Who are the villains, or at least the opposing sides, injecting these ideologies into bioethics?

On the one side are neoconservative and religiously oriented bioethicists. They are wary of where biomedicine and biotechnology are taking us. They speak in terms that are religious or quasi-religious. They have established their own journals, think tanks, and training programs. They have operated in the corridors of power both in the White House and in Congress. They are at ease with the Republican Party. They are backed by the deep pockets of very conservative foundations and wealthy philanthropists. They have no hesitancy in saying that they operate as bioethicists, but they rarely appear at the professional society meetings and do not publish often in the mainstream bioethics journals.

On the other side stand a larger but looser amalgam of left-liberal bioethicists. They are, on the whole, more at ease with the Democratic Party. They are also more at ease with science and technology then their conservative counterparts. While not always in love with every thought, proposal, experiment, or initiative emanating from the world of bioscience and technology, this group has no inherent fear or loathing of a scientific worldview. Indeed, they place their bets for a better tomorrow on scientific and technological progress. They speak primarily in secular terms drawn from philosophy or the law. Explicitly religious arguments make them nervous. They tend to dominate academia and the major bioethics programs located there. They run the mainstream journals, blogs, and training programs. With the significant and recent exception of the Center for American Progress, they don't have special access to left-wing foundations, think tanks, and philanthropists, but government, a few old-line foundations, and some corporations fund their work. They are a bit more nervous about admitting to being bioethicists in public places, despite the fact that their history is far longer than that of their conservative counterparts. When asked about ideology, they routinely swear that they are fair in their classrooms and that they present all points of view.

The divisions described above have been on display since the formation of the President's Council on Bioethics at the start of George W. Bush's presidency. Those on the left, secular end of bioethics had historically dominated most earlier federal bodies, such as the President's Commission to National Bioethics Advisory Commission. With the appointment of Leon Kass, MD, a new wind blew into Washington from the right. As the Kass-led President's Council on Bioethics did its work, issued its reports on various subjects, and replaced departing members with new individuals, the reactions to these activities tended to reveal the fault lines that had formed in bioethics. Secular liberals stewed, fretted, and griped; conservatives and religious bioethicists offered support and praise.

That there is a fight about what political stance bioethics should adopt toward biomedicine and biotechnology is beyond debate. Whether doing bioethics from an ideological point of view leads to sound bioethics is highly debatable. What is now beyond debate is the politicization of bioethics.

Some are dismayed by the level of discourse and rhetoric on display among bioethicists in talking to one another, and by how quickly things can turn nasty in public. A few figures in the field have tried engaging in quiet diplomacy between members of the two groups. They believe that it is inappropriate for those who style themselves students of ethics to engage in the fiery language associated with politics and ideology, and that politicization is fundamentally damaging to the future of bioethics. Others worry that the level of emotion characteristic of many current debates will simply disqualify bioethics and bioethicists from their historic and hard-earned role as independent voices of reason and as trustworthy sources of objective analysis in an America still deeply divided along religious, cultural, class, ethnic, and racial lines. They believe if bioethics cannot behave like a mature, thoughtful adult ought to, but rather insists, like some rebellious teenager, on throwing itself pell-mell into the melee of soundbite media conversation and the mud-slinging that passes for political debate in America, then bioethics is at best lost and at worst not worth keeping.

While many bemoan a loss in civil discourse, and while occasionally a foundation or the president of a university tries to do something to restore amity to the chatter of the polity, in general the amplification of rhetoric and sniping in bioethics reflects a deep schism in American society about the relative authority that ought to be accorded science and religion. In that sense, the politicization of bioethics may be unfortunate, but it is also unavoidable.

But it is not just the fact that bioethics issues became significant political issues over the past few decades. Bioethics is suffering mightily from its own success.

Since the late 1960s, bioethics has built a reputation as a valuable resource for health-care professionals, scientists, public policy makers, and patient advocacy groups. Bioethics has, despite the laments expressed in some recent scholarship about it in history and sociology, actually done some good: helping to build a system of human subjects protections for those involved in research, carving out more patient control over medical treatment, and laying out a framework to guide the procurement and allocation of cadaver organs and tissues. Bioethics has grown from a cottage industry of intellectually lonely misfits and malcontents eager to embrace others with common interests no matter their views or politics—the ethos that permeated the Hastings Center and the Kennedy Institute in the early 1970s—to a real field that whispers in the ears of presidents, issues rules to bind the inquiries of Nobel Prize winners, and is consulted by CEOs and the media for advice and analysis. Bioethics is a field that finds itself, unexpectedly, with power in the early years of the twenty-first century. Power brings with it not only new duties, responsibilities, and concerns but also, inevitably, politics.

Those who seek to shape policy and practice in biomedicine know that they can gain an edge by consorting with bioethicists. Lawyers looking for an edge for their clients in health- and science-related fields are now constantly on the prowl to find a bioethicist for their side. Companies and patient advocacy groups seek to better position themselves and protect their interests by engaging the services of bioethicists. Inevitably, this means that bioethics cannot feign an indifference to ideology or maintain a stance of studied neutrality in the face of controversies where the stakes are high and passions run deep.

I do not mean to say that efforts to turn down the rhetoric or to seek forums where thoughtful conversation and reflective dialogue are welcome ought not to be pursued. Nor would I maintain that the politicizing of bioethics is an excuse for every silly, stupid, incautious, or injudicious remark or thought. But I would argue that bioethics has taken a turn down a road from which there is no return. In asking to be a voice in the formation of policy, to be taken seriously in guiding the future of biomedical inquiry, to be heard by students in colleges, high schools, and professional schools, to place its new graduates in jobs on the staff of senators or in the offices of Fortune 500 companies, bioethics has made a bed it must now sleep in.

Bioethics has met power and can serve political purposes. It has made a difference, and it now wields power. No power exists in a political vacuum. The key to navigating the new world that bioethics finds itself in—the public arena, which is a stormy, unpredictable, and even dangerous place to be—is to admit these facts. Once it has conceded there is no turning away from them, bioethics will, like economics, political theory, and sociology before it, have to learn to live with power. One way to do so is to operate from explicit ideological perspectives. There may be other ways. But the days of hanging on the fringe and offering commentary solely in a prophetic mode are over. Bioethics is becoming more mature, and must now sometimes be done with a suit and tie.

Works Cited

Callahan, Daniel. 2003. Individual good and common good: A communitarian approach to bioethics. *Perspectives in Biology and Medicine* 46, no. 4: 496–507.

Caplan, Arthur L. 2005. Who lost China? A foreshadowing of today's ideological disputes in bioethics. *Hastings Center Report* 35, no. 3: 12–13.

Caplan, Arthur L. 2007. Rhetoric and reality in stem cell debates. *Society* 44, no. 4: 26–28.

Caplan, Arthur L. 2008. Bioethics grows up. *PLoS Biology* 6, no. 4: e95. http://biology.plosjournals.org.

Caplan, Arthur L., James J. McCartney, and Dominic A. Sisti, eds. 2006. *The Case of Terri Schiavo: Ethics at the End of Life*. Prometheus Books.

Cohen, Eric, and William Kristol. 2002. A clone by any other name. *Weekly Standard*, December 3.

Cohen, Eric, and William Kristol. 2004. The politics of bioethics: Playing defense is not enough. *Weekly Standard*, May 10.

Jonsen, Albert R., and Renee C. Fox. 1996. Bioethics, our crowd, and ideology. *Hastings Center Report* 26, no. 6: 3–7.

Kass, Leon. 2002. *Life, Liberty, and the Defense of Dignity*. Encounter Books.

Kristol, William, and Eric Cohen, eds. 2002. *The Future Is Now: America Confronts the New Genetics*. Rowman and Littlefield.

Smith, Wesley. 2000. *Culture of Death: The Assault on Medical Ethics in America*. Encounter Books.

12

A Catholic Progressive on Care and Conscience

Michael Rugnetta

Progressive bioethics does not simply promote scientific progress for its own sake while ignoring the question of what ends science should progress toward. Nor does progressive bioethics promote progress in science, technology, and biomedicine as an endless self-aggrandizing contest in which humanity must always try to best itself. Progressive bioethics promotes progress in these fields as a means of achieving expanded rights, freedom, and dignity for all human persons. To these ends, progressive bioethics endeavors to ensure not only expanded health-care options and access, but also free exercise of the individual conscience of both health-care workers and patients.

Conscience clauses—which preserve the right of health-care providers to refuse to provide a medical service because of a religious or moral objection—brings at least two progressive ends, freedom of conscience and the right of access to health-care options, in tension with each other. Moreover, the subject of conscience clauses, also called refusal clauses, brings into focus the necessity of determining which rights and obligations should take precedence when two such claims are in conflict. The subject is even more complicated when one considers the manner in which these rights and obligations are conferred upon different entities (individuals versus institutions), in different sectors (public versus private), and in different communities (religious versus civil society). Nevertheless, a careful sifting through of these myriad considerations reveals that the right of an individual health-care worker to freely exercise his or her conscience can be preserved while still protecting the reproductive rights of an individual patient in both the public and private sectors. These rights can also be accommodated both in civil society and in religious communities. An examination of the Roman Catholic approach to conscience clauses demonstrates a unique example of the latter.

As a political issue, the preservation and expansion of conscience clauses is of particular importance to many conservative Christian communities since it serves as another important battle in the culture wars over abortion, contraception, marriage, and sex. The Roman Catholic community poses a particular challenge on this issue not because there are individual Catholics who feel the need to exercise their consciences in one way or another, but because many Catholic health-care institutions exercise an institutional conscience along the lines of the official Church teaching, which strictly opposes essential reproductive services such as abortion, sterilization, contraception, and in vitro fertilization. Therefore, the official Catholic teaching on conscience clauses presents an opportunity to demonstrate not only how a progressive civil society should deal with that position, but also how other Catholic teachings which *are* progressive can theologically legitimize positions held by Catholics who disagree with the official Church position. Moreover, the issue of conscience clauses demonstrates how progressive Catholic teachings can enable the Catholic community as a whole—with its legitimate diversity of views—to cohere better with the larger civil society in which the Catholic community resides. First, however, it is instructive and necessary to discuss the historical and political background of conscience clauses in the United States.

The History and Politics of Conscience Clauses

Conscience clauses have gone through multiple permutations since they first appeared after the *Roe v. Wade* decision in 1973. Originally, these clauses sought to protect health-care workers or institutions that refused to participate in certain health-care practices (including the provision of contraceptives, sterilization, and abortion) on moral or religious grounds. In concrete terms, these health-care professionals and/or institutions are usually protected through guarantees of accreditation, government funding, employment, training, and/or education in spite of their refusals. However, if conscience-clause laws are to be both progressive and pragmatic, their ultimate goal should be to strike the right balance between the right of health-care professionals and/or institutions to provide care that falls in line with their moral or religious beliefs and the right of patients to have access to all forms of medical care.

The first conscience clause was enacted by Congress in 1973 immediately after *Roe v. Wade*, and is referred to as the Church Amendment (after Senator Frank Church). It said that the receipt of federal funds does not require an individual or entity to provide abortion and/or sterilization if it "would be contrary to [the individual's or entity's] religious beliefs or moral convictions" (42 U.S.C § 300a-7(b)). It also takes a "neutral stance" toward abortion and sterilization with regard to employment. In other words, an institution receiving federal funding may not discriminate in the hiring, firing, promotion, or the granting of privileges to physicians or staff members based on their performance of or their refusal to perform sterilization or abortion.

From the stipulations of the Church Amendment, we can see how conscience-clause laws attempt to balance freedom of conscience for both the provider and the patient. Ostensibly, the amendment's "neutral stance" respects the conscience of providers who agree to perform said services and those who refuse to perform them. However, there is no stipulation that ultimately guarantees the provision of abortion or sterilization procedures for a patient. According to Planned Parenthood, the American Public Health Association deems refusal clauses appropriate only if they provide an adequate plan for referral and do not disrupt or obstruct a patient's access to care with "untold delay or barrier." In order to accommodate the rights of the patient, Planned Parenthood recommends that professionals who refuse to provide a certain service do so consistently and inform their employers so that the proper arrangements can be made in a timely manner. This includes setting up a referral procedure. Pharmacists must direct the patient to a nearby facility that will provide the medication in question in a timely manner. If there is no way for a patient to obtain a medication by alternative means in a timely manner the refusing pharmacist is then required to dispense the medication (Preservation of Reproductive Health Care in Medicaid Managed Care 2003, cited in Planned Parenthood 2004, 3). These provisions strike a proper balance between the rights of service providers and patients much more pragmatically. Setting up a referral procedure in advance allows an institution to ensure that a patient will receive the service or medication that he or she is entitled to while also not demanding that the provider be actively complicit in its provision.

Since the passage of the Church Amendment, 46 states have passed laws that protect medical entities and certain professionals who "opt out" of administering abortions, 17 states protect doctors who refuse to perform sterilization, and 13 states allow some health-care providers to refuse to provide contraception-related services. Currently there are laws in Arkansas, Georgia, Mississippi, and South Dakota that specifically protect pharmacists who choose not to dispense emergency contraception. Colorado, Florida, Maine, and Tennessee have more general conscience-clause policies that do not mention pharmacists but probably would protect them. Illinois has a similar policy but also requires all pharmacies that stock contraceptives to dispense all forms of contraceptives. In California, refusal is allowed if the pharmacist's employer approves and the woman can still get the contraceptive in a timely manner (Guttmacher Institute 2007).

Evolution of Federal Conscience Clauses

Congress has passed numerous other measures since the mid 1990s to make sure that health-care professionals and insurance providers are not required to participate in abortions. Since 2004, Congress has attached the Hyde-Weldon Amendment to many appropriations bills. This amendment prevents government agencies from treating health-care providers differently if they refuse to perform or pay for abortions.

The Omnibus Consolidated Rescissions and Appropriations Act of 1996 prevented state and local governments from discriminating against health-care entities that refuse to undergo or provide abortion training, perform abortions, or provide referrals for abortions or abortion training (Feder 2005, 2). It is this last stipulation regarding referrals that can be regarded as an infringement on the conscience of the patient, since a lack of referral would hinder the patient from obtaining an abortion in a convenient and timely manner. Additionally, this law goes beyond being simply a conscience clause, since providers can deny service for any reason, not just or moral or religious grounds. This means that the decisions of health-care service providers are protected if they originate not only on the sturdy grounds of conscience but also, quite possibly, on the more shaky grounds of mere whimsy. In order to achieve a principled and pragmatic balance between the rights of the patient and the rights of the provider, the provider must exercise that right on the proper grounds of conscience—and nothing less—in order to hold the same amount of weight as a patient's right to the service that they need as prescribed by their physician.

An expanded form of the conscience clause occurred in the Balanced Budget Act of 1997. Congress's original intention was to keep managed-care plans from restricting the ability of doctors to discuss procedures with patients that the plan does not cover. However, this Act included an exception which allows managed-care plans funded by Medicare and by Medicaid to opt out of providing coverage or reimbursing providers for abortion services, referrals, or even counseling.

And starting in 2005, the Hyde-Weldon Amendment was attached to appropriations bills for the Department of Labor, the Department of Health and Human Services, and the Department of Education so that state and local governments could not deny federal funding to any "health care entity"—defined broadly to include health insurance companies and HMOs—if they refuse to perform, pay for, or refer for abortions (Feder 2005, 5).

Admittedly, the provisions in the three acts discussed above are much more difficult to evaluate than the provisions in the Church Amendment and state conscience clauses. This is because they to do not seek to accommodate an individual provider's conscience per se, but instead seek to accommodate the coverage decisions of government-funded health-care entities more broadly by allowing those entities to refuse coverage of abortion services, referrals, or counseling, as well as prohibiting the denial of funds to entities that refuse to perform, fund, or refer for abortions. This is a more understandable accommodation given the current state of America's health-care system, where it is—for better or worse—well within the purview of an insurance company or a health-maintenance organization to provide or deny coverage or reimbursement for a service on the grounds of efficient resource allocation. However, for an institution to go so far as to deny a referral or counseling—not just an actual medical service—seems to belie a health-care entity's ostensible desire to efficiently allocate resources and instead reveals a more genuine desire to impose a moral or religious view on patients. This would make the provision of government funds to such entities problematic, since such provision would encroach on the Establishment Clause of the First Amendment. Unfortunately, it would be extremely difficult to enforce a law that utilizes the motivation behind an entity's refusal—religious/moral vs. resource allocation—as the criterion for providing the entity with government funds. More generally, these federal provisions demonstrate the difficulty and ambiguity surrounding the government's funding of institutions that refuse to provide a controversial health-care service such as abortion, sterilization, or contraception. More important, it demonstrates why the most legitimate ground for allowing the refusal of a medical service is to protect an individual provider who is acting on his or her moral or religious conscience. When that allowance is expanded to accommodate entire health-care providing institutions, the moral grounds for such an allowance become much less legitimate.

The Official Catholic Position and Its Consequences

An easy assumption to make about the exercise of a religious conscience is that all adherents to a certain religion are of the same conscience. In the case of conscience clauses, the exercising of a religious conscience might seem to effectively mean that a true religious adherent might have no choice but to refuse to provide a reproductive health-care service if that is what his or her religious leaders demand. It might even seem that conscience clauses provide a way for a religious adherent to avoid having to choose between the religion and his or her obligation as a provider of a health-care service. However, Roman Catholicism illustrates how a religious adherent can find support in a faith's traditions and teaching to support a conscientious decision that does not fall in line with the official teaching of present-day church leaders but still be acting with a faithful conscience. This also demonstrates how a religion does not necessarily have to subjugate itself to the dictates of a progressive and religiously plural civil society in order to coexist with progressive values. Indeed, progressive values can guide religion from within. This allows for a productive dialogue between a progressive and religiously plural civil society and progressive religious communities, so that they can reassess their goals and principles for the sake of cooperating with one another. This process stands in stark contrast to simply assuming that religious communities will inevitably conflict with a progressive and religiously plural civil society and that one party or the other will have to make a trade-off in order to avoid a stalemate.

Refusal clauses have been strongly supported by both the U.S. Conference of Catholic Bishops and the Catholic Health Association. The CHA is the trade association of the Catholic health industry; it also represents the interests of Catholic health-care providers on Capitol Hill and in state legislatures. It claims to be "the largest health care provider in the nation," and it considers conscience clauses a major political priority.

Indeed, the fact that the CHA is "the largest health care provider in the nation" not only indicates the size of its political clout but also the ubiquity of its service in the public sphere. External research backs up this claim:

• According to the MergerWatch Project, four of the ten largest nonprofit health-care systems in the United States are Catholic institutions, including the largest (Catholics for a Free Choice 2005, 2).

• In 2003, there were more than 15.4 million emergency-room visits and more than 86 million outpatient visits to Catholic hospitals (ibid., 1).

• As of January 2005, there were 60 Catholic health-care systems present throughout all 50 states (ibid., 1).

• A 2002 study of nearly 600 religiously affiliated hospitals in the United States found that they received more than $45 billion in public funds. Approximately half of their revenues come from Medicare, Medicaid, and other government programs (ibid., 2).

• In many states, 30–40 percent of people who need emergency care visit a Catholic hospital (Catholics for a Free Choice 2002).

Although the private consciences of individual employees of Catholic hospitals and health-care systems should be respected, it is not obvious that the same applies to Catholic providers of public service. Their ubiquity in the public sphere and the fundamental nature of the services they provide places them under an obligation to play by public rules and not unnecessarily hinder their patients' access to care. Currently, however, Catholic hospitals are unnecessarily hindering their patients' access to care; and the numbers show it.

• A 1999 survey of 589 Catholic hospitals by CFFC revealed that 82 percent said they do not provide emergency contraceptives under any circumstances, even if a woman has been sexually assaulted (Catholics for a Free Choice 2005, 4).

• Only 22 percent provided a "meaningful referral"—i.e., continuity of care without an undue burden such as traveling long distances (ibid., 4).

• Only 9 percent provide emergency contraceptive services (ibid., 4).

• Only half of Catholic HMOs cover contraception or sterilization; abortion is largely unavailable (ibid., 4).

The Catholic Concept of Conscience and Its Role in Medical Ethics as It Relates to the Directives

Although publicly funded Catholic hospitals are obligated not to unnecessarily hinder their patients' access to care, this should not be seen as a demand that the Roman Catholic community suppress its religious conscience when engaging with civil society. Instead, this should be seen as a call for the Roman Catholic community to rediscover what it means for a faithful Roman Catholic to exercise his or her conscience. Indeed, some have considered conscience central to Catholic moral teaching. The centrality of conscience is derived from what Catholics deem to be God's greatest gift to human beings: free will. Only through the exercise of free will can an individual voluntarily choose God, goodness, and morality. Without free will and the ability to make free choices based on one's own conscience, it would be impossible for an individual to give meaning to his or her existence and actions.

For Roman Catholics, conscience is connected to the value of each human person. It has been described as an internal judge, the most noble audience for which an individual can act—essentially, conscience is that inviolable quality, intrinsic to man, that escapes any human interference (Maguire 1999, 32–33).

Yet in spite of this internal aspect of conscience—this desire to be true to the God in oneself—conscience must also be fully aware of how it affects external reality in order for it to be exercised ideally. In the long history of the Catholic moral tradition, this is referred to as the conflict between the subjective and the objective aspects of conscience. Subjectively, one's conscience can possess an intention that is either sincere or insincere. Objectively, one's conscience can possess information that is either true or erroneous (Curran 1999, 172).

Drawing on this classification, one's conscience can then take four forms. The ideal form is the true and sincere conscience; the worst form is the insincere and erroneous conscience. The other two forms are more ambiguous. However, the Catholic moral tradition does grant primacy to the subjective aspect of conscience and considers an act more questionable if it is true but insincere—e.g. donating money to help the poor just to impress people. In the fourth situation, in which the conscience is sincere in intention but based on erroneous information, one's error can be further be subdivided into two forms: vincible ignorance, where one is negligent or should have known better, and invincible ignorance, where ignorance is justifiable and one need not act with a guilty conscience (Curran 1999, 173).

Thomas Aquinas argued simply that one must follow an erroneous conscience, but did not address the issue of vincible versus invincible ignorance. However, he did emphasize that holding in contempt the dictates of an erroneous conscience is a mortal sin—even if that erroneous conscience contradicts the precepts of a superior (Curran 1999, 174).

Even at this early point in the development of Catholic moral philosophy, an emphasis on the *individual*—as opposed to the institutional or collective—nature of conscience was a crucial element of its definition. The progressive creativity of the Catholic tradition lies implicitly in that emphasis on the individual—or, more specifically, in the ongoing tension between the individual conscience and the individual's institutional superiors (both secular and religious). Indeed, for Catholics, there exists an evolutionary and revelatory aspect of God's will. The revelations that are said to exist in scripture constantly evolve as faithful human beings experience new situations. Yet there always remains that same tension between those who hold to the security of past answers and those who sense themselves called to find the answer that God seems to be demanding of them now within the complexities of their present situation in life. This calling is indeed a calling to a progressive Catholicism, one that can engage in dialogue with and cooperate with an ever-changing progressive civil society.

The progressive Catholic tradition can also be found in the writings of Saint Paul. In his letters, Paul grants a great degree of primacy to one's own conscience, yet he does not consider it to trump the conscience of others. This demonstrates how the individual Catholic can engage with and accommodate another individual who may not be a Catholic and may be of a different conscience on a given matter—and do so without harboring a guilty or false conscience. Indeed, Paul writes that "anything which does not arise from conviction is a sin" (Romans 14: 23). However, he also writes that sometimes it would be more loving to refrain from exercising one's own conscience in order to demonstrate respect for the conscience of another, even if that other's conscience is erroneous (Maguire 1999, 34). These teachings still apply today in discussions of the conscience clauses that are enacted to give pharmacists the right to deny emergency contraceptives to a patient on moral or religious grounds. A Catholic pharmacist need not defer to the official teaching of the Catholic Church and deny emergency contraceptives to a patient in order to be considered a faithful person acting in good conscience. In accordance with the teachings of Paul, it is permissible for a Catholic pharmacist—even one who strongly opposes contraception—to defer to the conscience of their patient or customer and provide it; and there is certainly enough permissibility for that pharmacist to give a patient or a customer a referral.

Aquinas writes that reason alone cannot give us faith (Maguire 1999, 38). The ultimate criterion for belief in God is the sense of peace that one experiences as one commits oneself to God in one's act of faith. It is the same way with conscience: Church teaching and the experience of others alone cannot resolve one's conscience. Certainly these things can be helpful, but ultimately all one has to depend on is the level of peace one feels when one commits to a particular choice.

In post-Reformation Catholicism, theologians taught that conscience could be guided but not tyrannized. They accepted the notion that the Church pronounced on the universal validity of general moral values but could not control the domain of personal conscience. From these teachings, Catholics were authorized to assess complex moral questions on the basis of their pros and cons, knowing that they could trust their subjective certainty. As Catholicism entered the age of the scientific revolution, it became more apparent that human beings had good reason to trust their own experience. Yet, as in the case of Galileo, the Church could still not accept that evidence might require it to examine its own teachings (Maguire 1999, 41).

In the wake of the First Vatican Council's 1870 decree on the infallibility of the papacy, John Henry Cardinal Newman built upon the idea of the evolutionary aspect of God's revelation in order to provide a more nuanced portrait of human conscience. Newman wrote that God implanted his Divine Law in all of his creatures. "This law," he asserted, "as apprehended in the minds of men, is called 'conscience'; and though it may suffer refraction in passing into the intellectual medium of each, it is not therefore so affected as to lose its character of being the Divine Law." (Maguire 1999, 43) Moreover, Newman quoted Aquinas's assertion that conscience is a "practical judgment . . . by which we judge what here and now is to be done as being good, or to be avoided as evil" (ibid, 45). This supports Newman's argument that the "Pope's infallibility . . . is engaged on general propositions, and in the condemnation of particular and given errors . . . a Pope is not infallible in his laws, nor in his administration, nor in his public policy" (ibid., 45). Clearly, we can see that nowadays most Catholics do in fact exercise their conscience against the Pope's public policy with respect to contraception: 75 percent of American Catholics believe that the church should allow contraception, and 98 percent of sexually active Catholic women say they have used a contraceptive method banned by the hierarchy (Limoges 2005, 36). This notion of the primacy of freedom of conscience in the face of official church teaching is backed up by multiple Catholic theological authorities:

Pope Pius XII: "Out of respect for those who are in good conscience, mistaken indeed, but invincibly so, and are of a different opinion, the church has felt herself prompted to act, and has acted, along the lines of tolerance." (D'Arcy 1961, 247)

Pope John XXIII: "We must never confuse error and the person who errs: (the latter) is always and above all a human being and retains in every case his dignity as a human person." (Pope John XXIII 1963)

Out of the Second Vatican Council came a renewed emphasis to engage in dialogue and embrace the social character of man in order to achieve new insights of truth. Indeed, neither the Official Church teaching nor the Directives of the U.S. Conference of Catholic Bishops is the final word.

Gaudium et Spes 16: "In loyalty to conscience, Christians unite with others in order to search for truth and to resolve, according to this truth, the many moral problems which arise in the life of individuals as well as in the life of society." (Pope Paul VI 1965)

Dignitatis Humanae 2: "The way in which the truth is sought must be in keeping with man's dignity and his social nature—that is by searching freely, with the help of instruction or education . . . through communication and dialogue." (Pope Paul VI 1965)

Sacramentum Mundi: "The judgment of the conscience is the ultimate definitive norm for the individual decision, but it does not thereby become a general norm for people faced with similar decisions." (Rahner 1968)

Humanae Vitae: "A right conscience is the true interpreter . . . of the objective moral order which was established by God." (Pope Paul VI 1968)

These teachings lend support to a pluralistic and democratic interaction of consciences—both within the Catholic community and within the pluralistic civil society of which members of the Catholic community are a part. An ideal legislative conscience clause would therefore take into account the differing faith traditions of patients and their right to not have a refusing pharmacist's conscience imposed on them. Moreover, it is in the interest of the employer to make sure that both employees' and customers' consciences are accommodated by having on staff pharmacists who will dispense contraceptives if one pharmacist refuses.

Finally, the directives of the U.S. Conference of Catholic Bishops, although strict and traditional when it comes to the denial of emergency contraceptives, still allow evidence of the physical world to influence its conscience in one specific circumstance. When a woman who has been a victim of sexual assault comes to a Catholic hospital, she is allowed to be given emergency contraceptives if it can be determined that fertilization has not taken place (U.S. Conference of Catholic Bishops 2008). The problem is that there is no test for the presence of a fertilized egg that can be given within the time frame that emergency contraception requires. As a result, practices among Catholic hospitals are inconsistent. The most frustrating fact about all of this is the U.S. Conference of Catholic Bishops' vincible ignorance of the documented evidence that progestin-only emergency contraceptive pills—such as the controversial over-the-counter Plan B—work only by preventing ovulation or fertilization and do not act as abortifacients or affect an already-fertilized embryo. Given this evidence, there is no excuse not to provide a woman who has been the victim of sexual assault with progestin-only emergency contraceptive pills. This situation clearly exemplifies a point made by J. F. Keenan SJ and Thomas R. Kopfensteiner: "When conscience is reduced simply to serving norms or an ideology, conscience is dead." (Maguire 1999, 52) Indeed, on the basis of progressive Catholic principles it can fairly be said that America's ever-broadening conscience clauses are in danger of eschewing evidence in order to serve an ideology, thereby—ironically—killing conscience. Furthermore, this is why institution-encompassing conscience clauses are far too broad. Although it is legitimate and admirable to recognize the right of individual medical professionals to decline to provide services they consider immoral, it nevertheless goes too far to grant to an entire federally funded institution—such as a hospital or a managed-care provider—the right to unduly hinder a patient's access to care by refusing to even refer the patient to another provider. Regardless of whatever allowances are made for the individual conscience of the pharmacist, institutions must not impose an ideology and instead defer to the individual conscience of the patient by respecting her or his right to comprehensive reproductive health care. This protection of and deference to the individual conscience of both the patient and the provider should be concretely realized through institutional efforts to accommodate both. The valuing of individual freedom of conscience over institutional ideology is one of the most elemental characteristics of progressive public policy. Furthermore, it should also be an elemental characteristic of any religion—such as Catholicism—that holds free will to be God's greatest gift.

Conclusion

Roman Catholic theology, although steeped in history and tradition, contains as much potential for progressivism as it does for conservatism. If Catholicism—and indeed any religion—desires equal treatment under the law in a religiously plural, democratic civil society, the progressive potential must be embraced. Indeed, it is not impossible for a religion to do so while also maintaining its eternal identity. An embrace of progressivism is the best way for a religion to achieve mutual compatibility with civil society. A religion does not need to make a false choice between the extremes of defying civil society and surrendering to secularization. A more transcendent path is possible—and that does not mean that religion and civil society must walk the tightrope of a strictly legalistic compromise. In a dynamic and progressive civil society, at any moment, in any situation, one kind of conscience might seek to assert itself—the pharmacist may be intransigent or the patient might be insistent—but ultimately it is the goal of public policy to ensure that certain safeguards are put in place so that these competing consciences balance out. Threats to freedom arise when one group of people are all of the same conscience and seek impose their views through a public policy that gives no recourse to people whose consciences oppose those views. On the issue of conscience clauses, ultimate safeguards such as a referral process or a previously established arrangement for another pharmacist to dispense the prescription must be in place. Most important, that is why institutions—especially institutions receiving public funding—must not mandate that all those under their purview submit to an "institutional conscience" without convenient recourse. Indeed, an institutional conscience is a contradiction in terms, but it is effectively what an institution imposes through its policies when recourse is not provided for those who disagree with those policies.

Although Roman Catholicism emphasizes hierarchy and institutional authority, it also maintains a foundation in the exercise of free will, which Catholics consider God's greatest gift. This negotiation between institutional authority and the individual conscience amidst the dynamic forces of social, political, and technological change give rise to new, creative interpretations of Catholic faith. Moreover, the respect that

Catholicism grants to the sincerity of one's conscience and the merciful deference it grants to the conscience of others represents a compassionate and progressive spiritual ethos to which all religions should aspire. It is an ethos that constantly orients itself toward the mutual accommodation of certain rights and obligations—even when they are seemingly in conflict. Indeed, it is an ethos to which all religions must aspire if they want to maintain relevance and meaning for faithful citizens of a religiously plural, democratic, civil society. Only then can humanity fully realize the promise of scientific and technological progress by achieving the ultimate ends of expanded rights, freedom, and dignity for all human persons of all consciences.

Acknowledgement

Parts of this chapter are taken from a Catholics for Choice pamphlet, *In Good Conscience—Respecting the Beliefs of Health-Care Providers and the Needs of Patients* (ISBN: 0-915365-88-X).

Works Cited

Catholics for a Free Choice. 2002. *Second Chance Denied: Emergency Contraception in Catholic Hospital Emergency Rooms.*

Catholics for a Free Choice. 2005. *Catholic Healthcare Update: The Facts about Catholic Healthcare in the United States.*

Curran, Charles E. 1999. *The Catholic Moral Tradition Today: A Synthesis.* Georgetown University Press.

D'Arcy, Eric. 1961. *Conscience and Its Right to Freedom.* Sheed and Ward.

Feder, Jody. 2005. The history and effect of abortion conscience clause laws. 109 Cong., 1st sess. S 1397, HR 3664. Government Printing Office. CRS Report for Congress.

Guttmacher Institute. 2007. *State Policies in Brief.*

Limoges, Roger J. 2005. Prescriptions denied. *Conscience* 26, no. 3: 36–38.

Maguire, John. 1999. *Conscience—a Cautionary Tale?* Church Archivists Press.

Planned Parenthood. 2004. Fact sheet: Refusal clauses: A threat to reproductive rights. http://www.plannedparenthood.org.

Pope John XXIII. 1963. *Pacem in terris.* http://www.vatican.va.

Pope Paul VI. 1965. *Dignitatis humanae.* http://www.vatican.va.

Pope Paul VI. 1965. *Gaudium et spes*. http://www.vatican.va.

Pope Paul VI. 1968. *Humanae vitae*. http://www.vatican.va.

Rahner, Karl, Ernst Cornelius, and Kevin Smyth. 1968. *Sacramentum mundi*, volume 1. Herder and Herder.

Rivera, Lourdes A., Ruth Roemer, and Lois Uttley. 2003. *Preservation of Reproductive Health Care in Medicaid Managed Care*. American Public Health Association.

Roe v. Wade. 410 U.S. 22 (1973).

"Sterilization or abortion." 42 U.S.C § 300a-7(B).

U.S. Conference of Catholic Bishops. 2008. *Ethical and Religious Directives for Catholic Health Care Services*, fourth edition. http://www.usccb.org.

13

Reforming Health Care: Ends and Means

Daniel Callahan

The United States is undergoing one of its periodic upheavals in health care. Every other developed country is having trouble with health care too. Whether we use the language of crisis or simply talk about the need for serious reform, there is general agreement that the U.S. cannot continue on its present course, nor can any other country. The universality of the problem shows that it would be a mistake to think that it is it is merely a matter of better organizing America's jerry-rigged system, not changing any underlying values (Hastings Center 1996).

Such a managerial move is necessary and needed, but the deeper problem is that no country has found a fully satisfactory way to manage the expensive and endless war against illness, disease, and death that is the modern medical enterprise. It is, perversely, a war that gets more, not less, expensive with every victory, every extra year of life gained. The healthier we get, the more we spend. Medical progress and equitable access to health care are on a collision course.

The reason for this conflict is not hard to find: aging societies, endlessly changing and improving technologies, and a rising baseline of public demand. Organizing health-care systems to cope with that dynamic has turned out to be daunting—and in the U.S. it is exacerbated by many conflicting interests and values specific to the American health-care system. But the problem is a common one. In modern medicine, with the full support of the public, we see an ambitious and insatiable agenda of expansive technological progress, one that seems designed to frustrate the devising of a reasonable, affordable, and sustainable health-care system.

By far the most important problem is that of formulating criteria for a humane health-care system, one that rests on a solid foundation of appropriate goals for health care. What, in short, is this enterprise of the pursuit of health, and of a good health-care system, all about (Schwartz 2005)? All other questions are subordinate to that overriding question. How one answers it will have much to do with the way the other issues are dealt with. Almost every ethical and social issue becomes a political issue, but health care is the most political, raising the widest range of issues, cutting through economics, public policy, law, local and national politics, history, and culture. And like the most serious political issues, it encounters the most serious ethical matters.

The devising and organizing of health-care systems has attracted only a small proportion of those working in ethics and social values. Much of the interest, moreover, has focused on one issue: just allocation of health-care resources. It is surely an important issue, but the scope of inquiry should be much broader. For many inside and outside of ethics, health care is thought to be "political" and thus not ethics at all—or to be the province mainly of other disciplines, notably law, political science, and economics. I think that is wrong. Aristotle held that ethics was a branch of politics, and he was right. Health care should be a central concern for those of us working in ethics, and those other disciplines need our help as much as we need theirs.

At the same time, if we are to talk about politics at all, much less about ethics and politics, there is a reality that must be acknowledged and dealt with. It is that the reform of health care, American or otherwise, is politically a complicated, messy matter. There are not, and there probably never will be, tidy political solutions, clearly realized moral principles (though many will be invoked), or any consistent and stable public consensus. The combination of ethics and politics in health care will not be advanced in the public interest if we are distracted by the recent ideological struggles in bioethics. They have been marked not just by disagreements (acceptable enough) but also by an unsavory degree of hostility and personal attacks. The issues are too complicated to be held hostage to that type of civic nastiness.

As a good indication of the depth of the pragmatic problem, two issues stand out for me. One of them is the consistent public opinion finding, going back many decades, that a majority of Americans support universal health care. A no less consistent finding is that they differ significantly on the means of achieving it, from market-dominated solutions to government solutions, and with many gradations in between. The other is how ineffective appeals to justice or a "right to health care" have been. They help to solve problems of ethical ends, but turn out to be of little help with the problem of political means, which is where America is stuck. Indignation about the uninsured is understandable and pertinent, but it does not much help in cutting through the underlying political and ideological struggles.

For those of us whose field is ethics, I believe, this situation poses a particular kind of puzzle and challenge. On the one hand, we need to bring to health-care ethics some set of principles and values to undergird whatever system seems to make the most sense to us, and may have some animating moral power. On the other, however, we no less need to know, in the name of political realities, how to find the limits to our ideologies and values, how and where and when prudence will dictate that we cut and trim, giving up a bit of our moralistic, rationalist souls to get the reform done. We must, that is, recall that we in ethics are seeking some universal values and goods, and at the same time imagine that we are in Congress, where a lot of ethics goes only a little way, not because legislators are immoral (at least not most of the time) but because it is their job to find ways to compromise, to bring off unpromising but necessary political marriages. The prize in this case will be serious and potentially efficacious reform, but we will not know in advance exactly what form it will take or whether it will serve our cherished principles.

Defining the Problem of Health Care

My aim here is to lay out a way of thinking about health care to make plausible my claim that it is the most important part of ethics. I also want to suggest some possible political directions for American health care.

Two Practical Issues: The Uninsured and Health-Care Costs
I believe there are two dominant practical issues (which I think of as "means") and two dominant foundational issues (which I call "ends"), but with considerable overlap.

I begin with the practical problems. One of them is the obvious need to provide health care for the 46 million Americans who do not have it, either temporarily or long term—a number that is growing steadily. I take that figure to be a moral scandal, both in the eyes of most Americans and even more in the eyes of every other developed country. I am hardly alone in calling for a universal health-care plan. The other practical problem is that of steadily rising health-care costs. Those costs are now increasing at a rate of 7 percent a year, a nasty kind of compound interest that will wreak havoc with the Medicare program in less than a decade and with the entire system not much later. We now spend $2.3 trillion per year on health care, and that is projected to rise to $4 trillion per year by 2015.

I call these issues practical not because they lack many ethical dimensions (although they do) but because they particularly challenge the future viability of American health care. To change that threatening trajectory in any significant way will require political ingenuity, complex organizational changes, and some way of reconciling a depressingly large set of competing interest groups.

Yet health-care costs are in many ways a more complex problem than universal care, particularly the cost of technologies. Hardly anyone thinks that having a large number of people uninsured is good, and public opinion polls have indicated for many years that the public favors some form of universal care. The prospect of seriously controlling technology costs is, however, not a happy one—necessary, perhaps, but disturbing. Such control threatens the constant innovation, and the prospect of saving future lives, that have given technology such a high public, commercial, and professional status. Everyone wants to get rid of useless technologies, but they are only part of the problem. Far worse is the prospect of limiting efficacious technologies. Doctor and patients love technology, industry makes a great deal of money from it, and the health-care system is designed primarily to provide it.

I have singled out cost and access as the most important practical problems, and others commonly add the improvement of quality to that list. I will stick to the first two only, and of course there is much agreement about their importance—but how they are handled and understood will be influenced by what are for me the most important foundational issues.

Two Foundational Issues: Ethical Roots and Medical Markets
One of the foundational issues is that of the ethical roots of health-care systems. Upon what understanding of human nature and ends should health-care systems rest? Upon what understanding of the quest for health and the ends of health care should they rest? Those are questions that American politicians are wary of confronting: they are too deep, too complicated, and politically more likely to drive people apart (politicians think) than to bring them together. Nonetheless, they are unavoidable.

My own approach is to start with the assumption—once uncontroversial but now more contentious—that we are finite creatures, born to live but eventually to die, whose life should be valued more for what is done with it than for how long it lasts. The aims of medicine should be to help us to have a good chance to go from being young to being old, to relieve us of our physical and mental disabilities if possible, to rehabilitate us as best it can if we are disabled, and to help us achieve as pain-free and peaceful a death as is possible. Medicine ought not to seek an indefinite extension of life, or aim to enhance our nature beyond the ordinary standards of good health, or to find medical ways of relieving us of all pain and suffering, many of which are now and always will be unavoidable.

I further believe that an average life expectancy of 80 or so, which Americans are approaching, is long enough to achieve most of the goods that life affords us if we are to achieve them at all. A longer life may of course add some benefits, and for many a longer life means a better life, but those individual benefits are not likely to be benefits for society. We may not like death and finitude, but they are good for us as a species, and necessary for the vitality of the community. I might add that, watching myself and my peers move into old age (I am now 77), I have seen no evidence of any special wisdom. Much as we were earlier in life, we are sometimes right and sometime wrong. I have been waiting for some great illumination, but it stubbornly refuses to arrive.

My assumptions are, I hope, sensible and prudent in themselves, but it is also difficult to realistically imagine that any more expansive assumptions will work to give the U.S. a just, accessible, and sustainable health-care system in the long run. The growing difficulties of the European health-care systems in controlling the cost of their otherwise economically efficient universal care should, along with America's problems, show that the United States is nearing the end of a managerial and organizational solution to the problem—other than a hazardous market solution such as pushing health care into the private sector. The present assumption of American health care—shared but not as intensively pursued in Europe—is that of the necessity of an open-ended pursuit of progress, positing no end or anything that would count as a final victory. More, just more.

There are, to be sure, many utopian voices claiming that more research will get us out of our present situation, that new technologies will lower costs, that we will find new ways of living in a transformed way, that we should never let up on progress. I have been hearing that kind of talk for all of my 40 years in health care, led by a drug and device industry that makes its money from innovation, by researchers who may believe it but also happen to get nice grants to pursue the next great breakthrough—and by many ordinary people who, understandably, do not like to get sick and die, and thus have a large stake in progress. But none of those hopes and dreams can do away with the nasty reality that, increasingly, we cannot pay for the open-ended progress we might like.

There is, of course, a connection between the two practical problems I began with and this fundamental one of a moral foundation for health care. If we are to have universal health care, to what should all of us be entitled? How far and in what ways should a health-care system be prepared to go in our quest for health and the avoidance of illness? If we are to control costs, to what extent and in what ways should cost be a consideration in patient care, including when it is needed to save a life?

The other fundamental problem is that of the tension between a belief that health care is best provided by the government versus a belief that it is best provided by the private sector. I will call this "the market problem" (Callahan and Wassuna 2006). This tension is obviously a part of the present debate over health-care reform. To call this a "tension" might seem either too strong or too mild. It is too strong if understood to be a black or white choice, either public or private. All sides seem to agree that some government support is necessary and that some private-sector care is desirable; getting the right balance is the real challenge. But it can seem too mild in light of the passionate and conflicting ideologies that mark the debate, usually cast in the most moralistic language possible: contrast the editorial pages of the *Washington Post* and the *Wall Street Journal*, or *The Nation* and the *Weekly Standard*.

I call this a foundational issue because, if one moves back from the inflamed rhetoric, there is a basic clash of values, each with a long history. That clash is between a communitarian understanding of our common life—man as a social animal—going back to Aristotle, and an understanding of man as a maximizer of self-interest—a more recent position, notably advanced by Adam Smith. Care must be taken here not to attribute cartoonish positions to these two thinkers. Aristotle surely understood the importance of the individual and of the need for virtuous individuals as the basis of a good society, just as Adam Smith understood the importance of empathy and a sustaining moral order as the basis of a good market society. Even so, they represent different conceptions of the individual and the community and offer divergent foundations for health care.

The appropriate starting point in thinking about the organization of health care is that of the foundational issues.

Solidarity and Universality

I favor universal health care because I see the preservation and pursuit of health as a necessity for the security and flourishing of society. Societies can surely survive with a high level of poor health, and many do. But a high mortality rate among mothers and children, common until hardly less than a century ago, was a misery of the first order. Hardly less so were the ravages of infectious and pandemic diseases. The prospect of living into old age was not, until recently, a good one, even when old age was thought to arrive earlier in the life span.

Medical progress, together with a better understanding of public health and the role of background social and economic conditions, made an enormous change in the perception of health and illness. Our health has steadily been brought under greater human control. We live longer and healthier lives than any humans in history. Nonetheless, we still live under the threat of sickness and death. They remain an enduring part of the human condition, and there is no reason to think it will ever be otherwise (Dubos 1983). This is true for each of us as an individual and for each society.

Since the threat is both individual and collective, a health-care system needs to be organized with that perception as its moral basis. For just that reason I have been much drawn to the European idea of solidarity as the foundational concept for the provision of health care (ter Meulen, Arts, and Muffels 2001). I take that term to mean that, in facing our finitude and mortality, we are all in it together. We share a common burden of illness, though to varying degrees, and a common threat of death. We thus need each other, and are joined by our shared fate. Moreover, when health care is enormously expensive, and getting more so all the time, then we face a common economic threat. Unless we are very affluent or wealthy, we cannot as individuals pay for our own care or that of our families. We need each other economically.

Using the language of rights or of justice to make the case for universal care has been common for decades (Daniels 2008). But the language of rights—a right to health care—has significantly faded, well symbolized by the President's Commission's rejection of that approach in 1983, replaced by an "obligation" on the part of government to provide care. Justice, however, has remained popular, but there are two reasons for wariness about using it to make the case for universal care. One of them is that, unless accompanied by a substantive account of what a health-care system ought to provide—to different age groups, and to those with different needs and conditions—it will remain abstract and of necessity procedural only. The other reason is related to the first: the concept of justice will not emotionally move the public or legislators to act, and, so far, has not done so in health care. I predict that not one of the political proposals for health reform will invoke it (though I will later in this chapter).

I press the notion of solidarity not because I rate its chance of acceptance as strong in American culture, but because the rights and justice approach has failed, and there can at least be no harm in trying a different route. Most important, it has served wonderfully in Europe (and Canada) to provide a sustained impetus for universal care. The ideology-driven characterization of universal care as "socialized medicine" is blithely ignorant of its European history, which goes back to the social teachings of Pope Leo XIII in the late nineteenth century and to the German Chancellor Otto von Bismarck during the same period. It was the latter who put in place the first universal health-care system, and who did so as a way of thwarting, rather than embracing, socialism.

The fact that every European government—right or left—since then gradually came to adopt that policy at the least should suggest that there is nothing inherently socialistic about it. Solidarity is first and foremost a moral concept, a way of characterizing the mutual obligations we should impose on ourselves for our common good, and that is the spirit of the European systems. In this country we understand that national defense at the federal level, and the provision of police and fire protection at the local level are necessary for our collective welfare. But not even the *Wall Street Journal* or the *Weekly Standard* refers to those government services as socialist. Why, one must ask Americans, is not our need for health care as important as our needs for national defense and for fire and police departments? Public opinion, for 50 years now, has favored universal care by a large majority, so it seems we get the point.

Choice and the Market

An increasing number of social conservatives now seem convinced that some kind of universal health-care system is needed; the main argument centers on the comparative role of market ideas and practices within such a system. Market-oriented advocates want a strong, even dominant role in the government-market balance. If solidarity is the central value for European-style universal care, it is choice and competition that are the central values for political conservatives (Porter and Teisberg 2006).

In one sense this emphasis can be seen as quintessentially a reflection of American individualism in its focus on choice, and a reflection of American business life in its focus on competition. Each individual should be free to choose his own kind of health care, and he should be able to select among a goodly range of commercial providers for its provision. But in another sense this emphasis no less reflects a history of an animus against government that can be traced back to Thomas

Jefferson, and more recently to the influence of Friedrich von Hayek and Milton Friedman.

One does not have to be a market enthusiast to appreciate the extent to which European health-care systems have worked in recent decades to bring many market practices (particularly competition) into their universal health-care systems. They do so, however, to improve and complement those systems, not to compete with them. Markets, I readily grant, can have a useful, but subordinate role to play in building a good health-care system.

But what moral weight ought we to give choice and competition? We can give them a light weight by recognizing that most people like to have a choice of doctors and hospitals and some say in the way they are treated, and are willing enough to have their care paid for through competing insurers. But many market advocates, at least on my reading, give choice and competition a heavier weight. They think that together choice and competition will not only lead to greater efficiency, but will be responding to one of the great moral imperatives of recent history. That imperative is, at the least, to save people from the tyranny and bureaucracy of government, but more important from going down what Hayek (1944 [1992]) called the "road to serfdom"—that is, the trajectory of socialism. Capitalism and the market need each other, joined together as two converging roads to preserve and promote human freedom.

The classical difficulty with that latter perspective is that the market is not a neutral tool, just as the "choice" that is part of its armamentarium is not neutral either. It may well be that the market does indeed promote national prosperity, but it does so by eschewing any judgment on the morality of the choices made. It aims to satisfy—but not judge—individual preferences. That makes it a dangerous solvent of traditions and moral values, a point well acknowledged by many social conservatives. That is why Irving Kristol (1978) could give capitalism only two cheers.

The provision of health care to a society, relying on the art and science of medicine as its primary vehicle, should be—as medicine itself should be—an altruistic enterprise, seeking the health of its citizens. To put at the heart of that enterprise a set of values that aims simply to bless free choice, not the goals of health care, and that makes individual choice, not health or the good of the community, its central commitment, is a dangerous move. Adam Smith believed in an invisible hand that could transmute self-interested motivations into a collective benefit. That may be true in many spheres of our civic and commercial lives, but it has yet to be demonstrated as true in health care.

Market proponents assume that, if letting individuals buy what they want from competing merchants is good for the sales of cell phones, automobiles, and TV sets, reducing their unit costs, it will likewise be good for health care. Kenneth Arrow (1963) presented a strong case that health care is different from the provision of other goods, given the disparity of knowledge between doctor and patient, and the uncertainty of treatments and outcomes.

But the real danger of choice and competition as basic values is that those choices, in American medicine, are heavily dictated by the commercial health-care industry, selling everything from insurance to drugs and devices (Relman 2007). Their motive is not health, though that can be good for profit, but the making of money for shareholders. American medicine is unique among the nations of the world for its thoroughgoing commercialization. Those industries have among the highest profits of all industries, aggressively sell their products (only the United States and New Zealand allow direct-to-consumer advertising), and no less aggressively spending millions of dollars to block any efforts to control that profit.

If "socialized medicine" is a bad idea, the high costs and comparatively poor health outcomes to match those costs make what can be called America's "commercialized medicine" far worse. If a nation divided cannot stand, neither can a health-care system that aims to join the historical altruism of medicine and the self-seeking of industry.

I will now return to my two practical problems.

Universal Care

I have not heard anyone argue that a large number of medically uninsured citizens is a good thing for American society. If nothing else, one way or another the public pays for their lack of care and for the economic and social burdens that their increased risk of illness and death carries with it. The 15,000 deaths a year that the Institute of Medicine has estimated as one of the prices Americans pay for that omission is not a trivial figure. Moreover, as a nation the United States can clearly afford to pay for universal care.

Why, then, does America not have it, particularly when public opinion polls have for so long found the public overwhelmingly in favor of it? One obvious answer is also found in those polls. There is considerable disagreement about what form universal care should take. It ranges along a continuum of government versus market oriented schemes, and there has always been some ambivalence about how much people are willing to pay to have it. Another reason is the sheer number of conflicting economic and professional interests, most with a stake in keeping the system as is, one that is profitable for doctors and for industry.

What I want to focus on here, however, is the deep-seated American hostility to government and a resistance to even looking carefully at the European health-care systems. By just about every possible standard, those systems are superior to the American health-care system. They have a greater life expectancy, lower mortality rates for infants, high-quality services that gain considerable popularity with the public, insurance coverage for everyone—and they do all that for considerably less money. By some major WHO and OECD studies, the United States ranks only 17th or 18th in overall comparisons of health care among developed countries.

Three mistakes are commonly made by Americans in their assessment of universal health-care systems. One of them is to think they are simply not worthwhile at all; they are government run or managed and thus ruled out a priori. The second is to constantly harp on the failures of two of those countries, Canada and the United Kingdom, as a generic indictment of them all. The third, related to the second, is a failure to recognize that there are two types of universal health-care systems: the so-called Beveridge systems, which are tax based and government run, and the Bismarckian Social Health Insurance (SHI) systems, which are financed by mandated employer and employee contributions and serviced by quasi-independent insurance programs.

The first point to be made is that the Canadian and U.K. tax-based systems are among the weakest of all the universal health-care systems, ranking down there near the United States on international surveys. It is those countries that have the notorious waiting lists, not to mention a serious shortage of doctors and nurses. But those failings are almost unique to the tax-based countries. The SHI systems have either no waiting lists or minor ones only, and all have better health outcomes at lower costs than the American system provides.

Those SHI systems—e.g., those of the Netherlands, France, and Switzerland—should particularly attract American attention. They offer some useful models for the United States. Among other things they offer a wide choice of physicians and other health services, competition among insurers, and coverage for all sectors of health care, including drugs and long-term care. Their lower costs are facilitated by strong government regulation, which typically encompasses the introduction and diffusion of new technologies, negotiations to set physician fees and hospital budgets, and cost controls on drug prices (Saltman, Busse, and Figueras 2004). But they have been willing to experiment with many forms of provider competition, to introduce managed competition, and to respond well to public demands for more consumer-directed health care. They do all this, including the regulatory impositions, in the service of sustaining universality, itself undergirded by a commitment to solidarity.

While most of the American health-care reform proposals now proliferating are (save for a few single-payer plans) hybrids of one kind or another—a mixture of public and private provision—few of those that rely on the private sector make any mention of government control of that sector. The European experience should, in that respect, teach the United States one important lesson: if costs are effectively to be controlled, the private sector cannot be allowed to run free. The U.S. experience with an unregulated private sector shows that it does not and cannot control costs, and neither consumer choice nor provider competition is likely to make more than a marginal difference.

Since I became interested in how the market interacts with medicine, well over a decade ago, I have been a diligent reader of the *Wall Street Journal*. Its editorial pages revel in knocking Canadian waiting lists, touting the glories of cheaper TV screens, assaulting socialized medicine, and glowingly describing the wonderful efficiencies and lower costs that the private sector can bring to health care. The best antidote to that kind of rhapsodizing are the news stories in its pages about the private sector, whose capacity for inefficiency, endless foul-ups, and unethical behavior is similar to the very worst of government behavior.

Medical Technology

I end with a discussion of medical technology. To my mind, it most vexingly joins together a practical and a foundational issue. As a practical matter, controlling the costs of technology is no less important than covering the uninsured. A 7 percent annual cost increase is already wreaking havoc with our present system and must be brought under control. By control I mean an annual cost increase that is no greater than the annual cost-of-living increase, in the range of 2–3 percent.

Health-care economists, in a rare moment of consensus, have determined that close to 50 percent of that increase comes from new technologies or the intensified use of old ones (Congressional Budget Office 2008). Any serious effort to control costs must, then, focus on the development and use of medical technology. If a hearty dislike of government is one of the obstacles to reform, the American infatuation with medical progress and technological innovation is one of the main obstacles to controlling its use and its costs.

The management of technology joins the foundational problem at the frontier of medical progress. That frontier, always moving, is where the present moment displays the success of past technological innovations in saving life and relieving suffering, and the promise that future innovations will do the same. We conquered polio and smallpox; now we search for cures to cancer, heart disease, and Alzheimer's Disease.

That we might stop with those earlier victories, and not try to bring comparable benefits to those dying of present diseases, is for most people simply a cruel and unthinkable idea. If it was right to bring beneficial innovation to earlier generations, it is no less right to bring it to present and future generations. The pharmaceutical industry has long made its case for the high prices it charges by invoking two related lines of defense. One of them is the hundreds of millions of dollars and the many years it takes to bring a new drug to market. The other is the large amounts of money needed to do the research that will save future lives—though just why those future lives outweigh the existing lives that could be saved by cheaper, more accessible drugs for present patients is never made clear.

Nonetheless, my guess is that such arguments in its defense resonate well with the public, and are readily heard being voiced by politicians. It is hard, moreover, to oppose the future lives argument, particularly if one has experienced the suffering and death of friends or family members from one of those many diseases not cured or controlled.

There are, then, many reasons to want to continue pressing forward on the frontier of progress. And there always will be, regardless of past progress. We are left, though, with the cost problem. There comes a point when continuing to cross that frontier, indifferent to the costs of doing so, is a basic threat to a fair distribution of present resources and no less so to future resources. This is a hard reality to accept, inviting denial, false hopes, and evasion.

My response is that if cost control is necessary for the good of the health-care system, and if the control of technology costs is a critical element in doing so, then these costs must be controlled. There will always be a frontier of medical progress and technological innovation. If everyone lived to 150 or 200, it would still be there. What's good for us at present, and what we need to live satisfactory lives, is not to win the endless war against illness and death, a war that cannot be won.

We need to learn how to live within the present frontiers of progress, moving ahead, if at all, slowly and carefully. If we don't get the ends right, the means will fail. If that seems a painful prescription for the health of health-care systems, there is a consoling thought: those of us in developed countries now live the longest and healthiest lives in human history. And we know that what most determines a nation's health status is not medical care but the general condition of life: jobs, education, environment, and income. Medical progress can bring us even better health, an obvious point I will not deny.

Yet at some point those additional benefits distort national budgets, take money from other urgent social needs, and elevate the idea of good health to a monstrous religion, consuming everything in its path. Even so, we will still get sick and die, but later rather than sooner, which I concede has something to be said for it. But maybe not that much: if I have learned anything in my aging, it is that later always arrives, and usually sooner than we expected.

Works Cited

Arrow, Kenneth J. 1963. Uncertainty and the welfare model of medical care. *American Economics Review* 53, no. 5: 941–973.

Callahan, Daniel, and Angela Wassuna. 2006. *Medicine and the Market: Equity v. Choice*. Johns Hopkins University Press.

Congressional Budget Office. 2008. Technological Change and the Growth of Health Care Spending.

Daniels, Norman. 2008. *Just Health*. Cambridge University Press.

Dubos, Rene. 1983. *Mirage of Health: Utopias, Progress, and Biological Change*. Basic Books.

Hastings Center. 1996. The goals of medicine: Setting new priorities. *Hastings Center Report* 25, no. 6: S-1–S-27.

Hayek, Friedrich A. 1944. *The Road to Serfdom*. University of Chicago Press [1992].

Kristol, Irving. 1978. *Two Cheers for Capitalism*. Basic Books.

Porter, Michael, and Elizabeth Olmstead Teisberg. 2006. *Redefining Health Care: Creating a Value-Based Competition on Results*. Harvard Business School Press.

Relman, Arnold. 2007. *A Second Opinion: Rescuing America's Health Care*. Public Affairs.

Saltman, Richard B., Reinhard Busse, and Josep Figueras, eds. 2004. *Social Health Insurance Systems in Western Europe*. Open University Press.

Schwartz, Aaron H., William B. Schultz, and Melissa Cox 2005. *Can We Say No? The Challenge of Rationing Health Care*. Brookings Institution Press.

ter Meulen, Rudd, Wil Arts, and Rudd Muffels, eds. 2001. *Solidarity in Health and Social Care in Europe*. Kluwer.

14

Finding Common Ground in Bioethics?

William F. May

Planners of the Bioethics Conference to be held at Albany, New York, in July 2006, invited me to talk in the closing session on the effort to find common ground in a field that had broken up into a series of Balkan states, clearly not so dismembered as old Yugoslavia but nevertheless a contested terrain. In the opening session of the conference, the planners also slated me to talk about the role of religion in public life, one of those cultural forces that has often proved broadly divisive in politics and specifically so in bioethics. This chapter will begin with the quest for common ground and close with the provocative question of religion's role in the field.

The Different Meanings of Common Ground in the Academy

The liberal arts college of the nineteenth century, largely Protestant in origin and small town in its setting, found its common ground in a general consensus on ends and goals that shaped its life. Reflecting this consensus, the president of the college often taught a course in ethics for all graduating seniors. This arrangement may have produced courses badly taught, but it also offered an important symbol. Ethics had not yet contracted into a merely technical subspecialty in religious or philosophical studies; it served rather to crown the student's education.

Various factors contributed to creating the possibility for this common ground. The college owed its existence and support to a specific religious tradition. Scripture supplied the institution with its uncontested canon. Moreover, even as the college distanced itself from its specific religious origin, the campus still offered a kind of common ground. Faculty members had not burrowed so deep into various areas of specialization as to lose their sense of themselves as a *collegium*. Further, although explicitly religious values lost their binding power to unify the college, faculty members informally agreed upon the "greats," the *auctores*, an acquaintance with whom defined the educated person. These authors constituted a quasi-religious sacred canopy under whose sheltering presence Western civilization would presumably continue to flourish.

Education was a matter of traditioning the young: handing on the past, honored as a valued thing on which to build. So went the ideal.

The twentieth-century positivist university no longer looked to substantive ends and goals to supply the institution with its common ground. Indeed, teachers were supposed to shed their values before entering the classroom in the disinterested pursuit of the truth. Values express only subjective, emotive preferences. They do not inhere in things; we read them into things. Spongy and subjective, values do not deserve a place at the lectern. Max Weber both described and recommended this restriction of the university to purely factual, objective inquiry in his lecture "Science as a Vocation." He asked, in effect, Do you want to be a leader or a teacher? a demagogue or a pedagogue? If you want to be a teacher or a pedagogue, then you must hang up your values on a peg in the cloakroom outside the classroom along with your hat and coat.

The ascetic restriction of the university to purely factual inquiry justified itself socially on the grounds that the acquisition of objective knowledge eventually produces skilled persons. Sold in the marketplace, these skills would enable students to enter into the slipstream of modern power. Ultimately, the objectivist creed of the faculty encouraged careerism in students. Faculty members could not pose questions of value in the classroom without descending into advocacy. While teachers can transmit a knowledge-based power, they cannot raise questions about its responsible uses. Ends cannot claim objective public status. Thus graduates can treat their knowledge as a purely private possession, which they may manage as they please in the pursuit of their careers.

However, while the university expanded and divided into a large array of disciplines and professional schools, it still offered a place to find common ground, in the form of a training ground. It prepared students for entry into a rich array of careers that would open doors to a more promising future than was available to their parents on the family farm or who had emigrated from Europe.

Education no longer oriented students to a valued past, but pointed them toward, and equipped them for, a better future.

The counterculture movement of the late 1960s and the early 1970s oriented neither to past tradition nor to future prospects, but to the present. Journalists dubbed those under 30 at the time the "Now Generation." The Vietnam War discredited the national future, proffered under the slogan "light at the end of the tunnel." The signature film of the period, *The Graduate*, exposed the tawdriness of any customary, personal future: when the drunken friend of young Benjamin Braddock's father urged the graduate to get into plastics; he offered, in effect, a plastic promised land.

At the Albany Conference on Bioethics, Dr. Edmund Pellegrino noted that the countercultural movement emerged as a general protest against authority—parental, professional, and bureaucratic—and that bioethics as a self-conscious discipline developed in this period. It was no accident that bioethics rode into town on the principle of autonomy, posing questions about uncritically accepted professional authority (Pellegrino 2006).

However, the revolutionary scope and force of the counterculture should not be exaggerated. When its force was spent, it left the liberal arts college and the research university, with its brace of professional schools, still standing.

Bioethics

The field of bioethics rose in part because the professions of medicine and nursing—in their successes as well as their failures—generated moral problems which professional training of itself did not solve. Issues such as confidentiality and truth-telling, care at the beginning and the end of life, experimentation on human subjects, high-risk therapies, the fair distribution of the scarce good of health care, and the deployment and quality control of professional services inspired ethicists to write elegant articles and deliver helpful lectures at grand rounds and in classroom sessions. They made available a vocabulary and a repair kit useful in equipping young professionals to deal with problems that surfaced on the training ground of the medical school and the teaching hospital. I do not mean by the image of a repair kit to diminish the significance but to specify the kind of ethics offered: ethics as problem-solving. The kit furnished basic philosophical resources that helped clarify and advance the decisions doctors needed to make. To that end, a literature emerged that drew on a limited canon—Kant vs. Mill, Nozick vs. Rawls—that marked out the parameters within which discussion often took place: deontologists vs. utilitarians, libertarians vs. egalitarians.

The debates were sharp and often adversarial, but opponents were recognizable and locatable within acknowledged boundaries. They could work together on national commissions. Indeed, it was even possible for a distinguished descendent of the deontological camp, James Childress (by way of Kant and W. D. Ross), to partner with an equally impressive descendent of the utilitarians, Tom Beauchamp (by way of David Hume), and produce the classic volume in the field, *Principles of Biomedical Ethics*. These co-authors proposed the four principles of non-maleficence, beneficence, justice, and respect for autonomy, which came to be known as the Georgetown Mantra, to guide reflection in the field. For varying reasons, these leaders resisted the term "principlist" to characterize themselves, but the basic impact of their work was to identify principles as the common ground for the field, which tended to push contending religious and philosophical positions (both their own and others) into the background.

Bioethics also developed a common ground for its work in the form of its pedagogical method. It relied heavily on the case method of study that already prevailed not only in medical schools, but also in schools of law and business. And it set out to engage in the important task of solving problems that the very success of medical practice generated, but a purely technical education did not retire.

Belonging to a subspecies of applied ethics, bioethicists have often found themselves under attack from two sides. On one side, some theorists in philosophy dismiss applied ethicists as derivative, unoriginal, and parasitically dependent upon foundational work which they themselves have not done. From the other side, practitioners in medicine, law, and business find them insufficiently experienced in the field to which they would apply their judgments. Applied ethicists seem to occupy a no-man's land, carrying water from wells which they have not dug to fight fires which they cannot quite find. Such is the unfriendly fire to which academics expose themselves when they cross borders as applied ethicists.

The chair of the President's Council on Bioethics, Dr. Leon Kass, threw down a different challenge to the field, when, under presidential appointment, he helped recruit a Council on Bioethics in the late fall of 2001 that would address such issues as cloning and genetic enhancement. Kass believed that those issues, among other scientific undertakings, pose basic questions about the human condition and prospect, specifically about human parenting (and about parenting the human future itself through the trajectories of scientific achievement), that cannot be adequately recognized and addressed through incrementalist problem-solving.

Kass announced that he did not see it as a primary goal of the Council to achieve unanimity on discrete public policies. That goal would require either stacking in advance the membership of the Council to guarantee consensus or producing a common denominator document, perhaps acceptable to a more diverse Council membership, but so vacant as to be unhelpful. Instead, he hoped that the Council might draft reports that all members of the Council could sign as stating the arguments on both sides of issues as fairly and reflectively as possible. Only secondarily, where possible, would the documents record where individual members stood. If successful in this approach, the Council would contribute to public culture, to the national public conversation on such issues, even if it did not directly influence public policy with the clout of unanimity.

As bioethicists well know, all members of the Council signed the first document on cloning, as stating the arguments pro and con; all members also registered their individual votes in favor of a ban on cloning children. However, the Council divided, ten members in favor and seven opposed, on a moratorium on federal funding for cloning solely for biomedical research. The seven opposed, myself included, voted in favor of the research, with firm regulations governing both federal and privately funded research.

Critics pointed to the small number of active physician-scientists on the Council (six out of the original 18 members) and to the presence of some members linked with the pro-life movement. Some critics also linked religious belief exclusively with the latter movement. However, as far as I know, the only ordained minister on the Council (myself) and at least two other persons of strong religious convictions joined the minority opposition to the moratorium on federally funded research. The makeup of the original Council was more complex than its critics supposed. Nevertheless, the chair's ambitious agenda to explore the human condition and prospect and call for a "richer bioethics" threw down a challenge to the previously defined common ground for bioethics. Furthering that challenge, Kass directed the staff to develop and publish *Being Human*, an anthology that offered a rich treasury of short stories, film scripts, poems, meditative pieces, excerpts from novels, and the like, intended to enlarge the canon in the field. The volume teaches beautifully, I should add, but it also includes nothing by Mill, Nozick, and Rawls, and only five pages from Kant's *Fundamental Principles of the Metaphysics of Morals*. Although Kass admirably sought to enrich bioethics, this deletion, in effect, excluded an important range of literature and, by implication, tended to marginalize questions of patient autonomy and distributive justice. On the latter issue, for example, some conservative members of the President's Council dismissed the question of universal access to health care as a budgetary issue, not a moral issue. Only much later in the deliberations of the Council, in a volume titled *Taking Care* that focused on the problem of long-term care for Alzheimer's patients, did a Council document recognize that access to health care rose to the level of a moral issue (President's Council on Bioethics 2005).

The series of documents from the President's Council (six volumes in three years) and reactions to those documents (and to changes in the roster of Council members) made it clear that the field had broken up into rival traditions with differing views of what constituted the field.

The Common Ground as a "Campus"

How can bioethics continue as a field of study, inasmuch as it has broken up into differing traditions, especially in a university that can no longer define itself as shaped by a single tradition (in the manner of the old liberal arts college) or that can no longer claim the commanding heights of a traditionless transmission of knowledge, which students convert as they please into careers? How do we conceive of a university whose common ground is neither a sacred canopy nor a training ground, few questions asked, for eventual high-flyers in the professions?

Perhaps 'campus' is an apt term for describing our common ground, both as a field and as a university. The Latin root of 'campus' refers literally to an open field, a flat place, the traditional site where rival armies encamp and fight. The university provides one of those precious open spaces in a civilization where the deep cultural conflicts within it can surface. Just as the legal system permits us to substitute a contest in the courts for a brawl in the streets in order that justice be done, so the campus allows us to marshal arguments, rather than troops, in order that our common life in the search for truth may be pursued and prized.

Engaging one another within and across traditions—attentively, civilly, and thoughtfully—means that a given tradition in the university does not simply perpetuate itself umbilically. Contest and dialogue with other traditions opens up a cognitive distance that allows us to visit a received tradition afresh. The aliveness of a tradition depends upon this re-visiting. At the same time, this revisiting also may entail a re-visioning of the world that a given range of human practice presents. This re-visioning creates some cognitive distance and space between practitioners and their field of practice, and thus invites them to act in fresh ways. Corrective vision of this sort offers an immensely practical freedom. We cannot change our behavior unless, in some respects, our perception of the world also changes. In this task, the ethicist theorizes—literally—and thus helps liberate.

Unfortunately, theorizing suggests to the practical person a remote and abstract enterprise, lacking in relevance and payoff, blindly distant from the world of practice. But classically understood, the theorist provides a fresh and liberating vision of the world. The word 'theory' in its Greek root refers to vision. Appropriately, the word 'theater' is also related to *theoria*, because theater, like good theory, presents us with a world to see and frees us from the local and the given. Thus applied ethics has a theoretical component related to liberating insight and vision. So conceived, moral reflection does not merely scan the world as it is or prepare leaders for the professions as they are. Rather it entails a knowledgeable re-visioning of foundations and ends. Through this cognitive illumination, it serves, in some limited way, the human capacity for resolution and decision. Bioethicists may not always eliminate moral quandaries and none has drafted a perfect health-care plan. The humanities do not bake bread. They contribute to public education and culture more often than directly to public policy. But the humanities open up a wider horizon in which quandaries and systems may be seen for what they are and thus become other than they were. The humanities help correct our perceptions of the world as it appears to the myopia of accepted practice and the astigmatism of vice. To this degree, they help us to serve the worlds in which we work, not perfectly, but better.

The Role of Religion in Bioethics

The difficulty of finding common ground in bioethics surfaced long before the controversies surrounding the reports of the President's Council on Bioethics and the ideological makeup of its members. I had a taste of it years earlier as a member of the subgroup on Ethical Foundations for the Clinton Task Force on Health Care Reform. Members of the subgroup were all committed to the importance of reform. However, the subgroup split into two approaches to justifying access to health care. The philosophers wanted a declaration of principles that rose above the particularities of any and all religious traditions. Only by shedding specific religious and cultural traditions could one arrive at common ground; that is, at the universal, the foundational, and therefore at the truly persuasive and compelling.

Meanwhile, other members of the subgroup, including the physician Steven Miles and myself, were disposed, both intellectually and politically, to include appeals to traditions, cultural and religious. We believed that access to health care, as well as to other fundamental goods, such as food, clothing, and shelter, were discernibly part of the national covenant to promote the general welfare.

Some secular participants in the debate (and some religious participants) favored removing religious language from the discussion altogether. They appealed to the principle of the separation of church and state (a phrase not found in the Constitution, but often linked with the

First Amendment). However, they overlooked an important feature of the First Amendment. Undoubtedly, it prohibits the enactment of laws for purely religious purposes, such as the establishment or advancement of a particular religious tradition. It hardly intends, however, to deny to citizens religious motives and reasons for believing and acting and voting as they do on matters of public purpose and the common good. Nor does it deny them the right to explain those motives and reasons to others. Religious liberty is not a purely private liberty. Otherwise it would not be clustered with the freedoms of speech, assembly, and press, all of which, together, supply the means by which people can assemble, speak, write, and reach their judgments on matters of great moment and the common good.

Other secular critics feared that religious language is inherently divisive and therefore unsuitable for discourse about common political purpose. They have a point. Consider, for example, the words of Rod Parsely, pastor of a 12,000-member church on the outskirts of Columbus, Ohio (Fitzgerald 2006, 27), as he cried out to a 1,000 people on the steps of the statehouse: "Sound an alarm! . . . A Holy Ghost invasion is taking place. Man your battle stations, ready your weapons, lock and load!" Or again, in his Sunday worship service, Parsely "called for a spiritual army to 'track down our adversary, defeat him valiantly, then stand upon his carcass'" (2006, 29). In such outbursts, religious leaders, like Pastor Parsely, confuse biblical monotheism with Manichaean dualism.

Any serious attempt to address Americans politically as they vote on questions of the common good will need to clarify the differences between monotheism and the lava flow of religious dualism that would flood the ballot box, the legislative halls, and the courts. However, Americans will not heal religious ravings and divisiveness simply by suppressing appeals to religion. Lincoln knew this when he called upon incandescent religious language to help the nation recover its badly shattered unity toward the end of the Civil War. In Lincoln's time and our own, we misread the politics of the nation if we assume that we are indeterminately a public, religious or secular. Rather the health of the nation depends upon recognizing that in forging national policies, we must appeal to diverse publics within the public at large, among them, religious publics with their own particular memories, beliefs, and motives for action which bear on questions of common purpose.

Some religious thinkers, as well as some secular thinkers, are reluctant to see religious leaders engaged in public issues. Many ministers in the mainline Protestant churches (if I may focus on the traditions I know best) hold back on political issues, not from the secular fear of the divisiveness of religion, but because of a distaste for the inevitable divisiveness of politics. For the sake of harmony, keep mum on politics. It spoils community. Too many prophets rattling about will undercut the ability of ministers and priests to fulfill their pastoral duties. Ministers need to be able to marry and bury and console Republicans as well as Democrats, communitarians as well as libertarians, doves as well as hawks. Effective pastoring, so the argument goes, needs to muzzle prophetic judgments against the political order, except in response to egregious wrongs, such as Nazi genocide policies. Otherwise, the church should keep its voice out of politics and the market place altogether. This approach, in my judgment, insufficiently honors the degree to which the gospel itself, while not supplying detailed blueprints for political and economic institutions, does furnish us narratively with goals by which our imperfect work in the world of politics should be measured. Securing access to the healing pool is one such issue. Care of the sick, the imprisoned, the poor, and the vulnerable is a recurrent theme in scripture. The gospel of John, for example, recounts the plight of a sick man lying near the healing pool of Bethzatha without access to the pool. So near and yet so far. No American Christian can read this story, among so many others on the subject of healing, and duck the vexing question of access. America has a healing pool the likes of which the world has never seen, from Johns Hopkins to the humblest of local community hospitals. Yet we face a huge barrier to the healing pool in this age of turnstile medicine. One out of every seven Americans has no insurance ticket through the turnstile; many more are underinsured. We are the only industrial nation that fails to offer health care to all its citizens.

While the passage from John exposes to shame the country's neglect of access to health care, neither the churches nor scripture offer an embossed blueprint for a health-care system. A fixed Christian blueprint would leave too little room for good-faith differences between people of faith, both on principles and their relative weight. It would also leave too little room for subsequent moral criticism of a particular design and its realization; and it might wrongly imply that only Christians could arrive at foundations for reform. One need not hold to the Christian faith to affirm health care as a fundamental public good among other goods. Monotheists need not always feel obliged to forge distinctive policies on any and all issues to prove God's uniqueness, as the country goes about the awesome business of wielding power for the common good.

Protestant sectarians have defined the role of religion in politics differently from the mainline churches. Sectarians divide into two basic reactions: either they despair of the political order altogether and withdraw from politics entirely to form a distinctive Christian culture cut off from the society at large; or they launch an occasional crusade to purify a fallen world. The latter, activist response has gained more currency in recent American history, often producing a single-issue vehemence in politics. It has led to targeted campaigns against slavery and civil rights on the left and against communism, abortion, stem cell research, and gay marriage on the right. Sectarian (and other) churches given to a single issue politics measure every politician—up or down—on that issue alone. On this issue, the church speaks with a deafening voice. On other issues (such as, for example, access to health care), it is silent. The reason for this silence traces, in part, to a conception of religious ethics, which identifies it solely with unexceptionable rules. The church should venture into politics only when it is a matter of enforcing universally valid rules. Other decisions are budgetary and political; they do not rise to the level of ethics. As I discussed earlier, some members of the current President's Council on Bioethics argued that the use of pre-implanted embryos in research is a moral issue that should concern the nation; however, the issue of access to health care (the concern of the earlier Clinton Task Force on Health Care Reform) is merely a budgetary and political issue on which moralists and the church should remain silent.

Two competing metaphors seem to underlie this understanding of religious ethics and the role of the church in politics. On the one hand, absolutists, among the sectarians (and others), are trumpers. They associate a moral principle with the unexceptionable, the universally valid. Trumpers see their opponents as weighers and balancers in the moral life who lack moral gravity. They are not principle-oriented in their ethics. Trumpers dismiss weighers and balancers as relativists. They worry that unless one holds to a principle as unexceptionable, one slides down the slope into the relative.

Absolutists do not leave room for those who are pluralists in their thinking about ethics and politics. Pluralists recognize that principles are multiple and therefore may call for the difficult task of weighing and balancing when they conflict. Such pluralism differs from both relativism and absolutism, perhaps more than absolutism and relativism differ from one another.

On the face of it, absolutists and relativists seem irreconcilably opposed. Absolutists argue that the validity of a principle depends upon its universality (there can be no exceptions). Relativists have discovered exceptions, and therefore deny the principle.

At a deeper level, however, absolutists and relativists agree. Both subscribe to the view that the validity of a principle depends upon its universality, its heft depends upon its overridingness in all circumstances. Neither has discovered the wisdom of the ancient observation that moral principles/laws are true for the most part. "For the most part" is not a loophole in and through which one can escape into the sea of the relative. They are true for the most part in the sense that they reach their territorial limit in those cases where they must yield to another principle or good, the more stringent, the more urgent in the particular case. Further, even in that yielding, the original principle does not simply vanish; it maintains pressure upon one, in how one pursues the supervening good; and it generates, in some cases, duties of reparation and gratitude—for losses imposed and benefits received.

Such is the rough landscape of policy making, in which one may need to compromise not in the sense of defecting from duty, but honoring principles which are multiple.

On these matters, I happen to differ from my college teacher in Christian Ethics, Paul Ramsey, who believed that a Christian ethic of love should embody itself only in exceptionless moral rules that "build a floor" under the moral life (1967). The image of "building a floor" is telling. The metaphor inadequately describes the full range of the moral life; ethics also calls for a building above. Indeed, the etymological root of the word *ethos* suggests as much. *Ethos* traces back to the dwelling place of an animal. An animal needs a habitat, a sheltering arrangement of space that accommodates its varying needs and movements; and human beings, naked apes, the most extraordinary, complicated, and vulnerable of creatures, need for their dwelling a full range of customs and practices and ways of being that protect and support not only human survival, but flourishing.

A floor of firm, unexceptionable rules does not fully describe what humans need. They also need a building that anticipates the range of activities and goods to be sustained in the human household. Decisions about the allocations of space, balance, and proportion in response to changing needs reflect the kind of beings we are, the overriding ends to which we aspire, and the goods and burdens which of necessity and by grace we may be called upon to share. In my judgment, the majority on Kass's President's Council on Bioethics focused primarily on the question of stem cell research because they felt that the debate on that issue could be formulated in terms of an unexceptionable rule, but they neglected almost entirely the question of access to health care under a ramshackle, wasteful system that bars access to so many.

Conclusion

In this chapter, I have restricted myself largely to some of the good-faith differences and debates within the Christian tradition, but a full tour of the current religious scene would have had to deal with the most powerful movement across the last twenty-five years—the religious right— whose reach today has extended into a huge range of domestic and foreign policy issues.

Let me close simply by locating this movement religiously. In my judgment, the religious right is not itself a version of Christian monotheism. While it poaches on biblical language, it is a competing, rival religion; it is religious dualism, a dualism, which, in varying ways afflicts each of the Abrahamic, monotheistic traditions of the West—Judaism, Christianity, and Islam.

Dualists differ from monotheists in that they recognize two ultimates, not one. They make the Devil coequal to God. They are metaphysically grim; they organize their lives and institutions obsessively around the struggle against satanic power. Second, they are morally blind. They identify themselves with God and their enemies and competitors with Satan. They split the human race into two, an ultimate denial of common ground. They have lost the sense that all—without exception—fall short of the glory of God. Such dualism moves toward the collapse of politics. It instinctively opts for military confrontation rather than the untidiness of political compromise. It is wary of any and all loss of control, whether through the entanglements of international agreement or through checks and balances that constrain from within. This dualist mindset has influenced our current national life far more than intramural debates within Christianity about the relations of the church to public life, of which issues in bioethics are but a small piece.

I have not called this movement, as it is usually described in the media, the "Christian right." Strictly speaking, its leaders urge upon the country not Christianity, but a different religion: dualism, not monotheism. Saint Augustine was the founding theologian in the West who recognized what was at stake here. In his great treatise against the Manichaean heresy of dualism, he did not locate dualists on a spectrum of Christian left, center, and right, as though he were describing simply different colorations of the same thing.

The metaphor of a spectrum, which the modern media use when they refer to a Christian left, center, and right, assumes a single beam of light which refracts into different colors. The Manichaeans, Augustine recognized, offered a different beam of light altogether, not simply an alternative shade of Christian monotheism. The Manichaean dualists were not a Christian right, they were something different; and, they were wrong. However, Augustine, once a Manichaean himself, recognized the lure of dualism in us all. He also declared that "to heal heretics is better than to destroy them." God's will is "that they should be amended rather than destroyed. And in every case . . . we must believe that the designed effect is the healing of men, and not their ruin." (Roberts et al. 1994, 29) In effect, Augustine offered a warning to monotheists on the current scene. They will themselves fall into their own peculiar Manichaeism if they simply slam the door shut on the dualists and create a reflexive dualism of their own making—identifying the bad guys with the current, ascendant dualists and the good guys with the dwindling band of mono-theists in the mainline churches. Believers in God as the Alpha and Omega owe God a more hopeful, open, and confident politics than that. They must persist in engaging with all sorts and conditions of men and women—libertarians, egalitarians, and communitarians, dualists and secularists, and religionists of every stripe—in the open terrain of the academy, and contribute as they can in building humane and habitable institutions in the rough terrain of modern politics. A rough terrain it is, but it is our common ground.

Works Cited

Augustine. 1994. Against the epistle of Manicaeus called fundamental. In *The Nicene and Post-Nicene Fathers*, first series, volume 4, ed. A. Roberts et al. Hendrickson.

Beauchamp, Tom, and James Childress. 1994. *Principles of Biomedical Ethics*, fourth edition. Oxford University Press.

Fitzgerald, Frances. 2006. Holy Toledo. *The New Yorker*, July 31.

Kant, Immanuel. 1949. *Fundamental Principles of the Metaphysics of Morals*, ed. T. Abbott. Prentice-Hall.

Pellegrino, Edmund. 2006. The inaugural John A. Balint lecture. Presented at conference titled Bioethics And Politics: The Future of Bioethics in a Divided Democracy, Albany, New York.

President's Council on Bioethics. 2005. *Taking Care: Ethical Caregiving in Our Aging Society*.

Ramsey, Paul. 1967. *Deeds and Rules in Christian Ethics*. Scribner.

Afterword

Sam Berger and Jonathan D. Moreno

Since we undertook this project, there have been great changes in the political and economic spheres, presenting new opportunities and challenges for progressives. Most significant has been the alteration of the political landscape. What began in 2006 as a shift away from conservatism became a historic victory for progressives in 2008; progressives won not only the presidency, but also large majorities in the House and the Senate. The most significant change has been the election of President Barack Obama. In 2008 the voters firmly repudiated the Bush administration in favor of a candidate who campaigned on progressive values that have been echoed throughout the articles in this volume: focusing on real-world results, relying on scientific and technical expertise and fact, and making government more responsive to everyday Americans. Whether this support will be transformed into a new era of progressive government remains to be seen. Surely much will depend on the success or failure of the Obama presidency.

We have also witnessed the onset of one of the worst financial, and now economic, crises in a hundred years, which severely threatens the financial industry, the savings and pensions of millions of Americans, and the long-term health of other sectors of our economy. The failures of deregulation and over-reliance on market capitalism have been starkly demonstrated. President Obama and Congress will have their hands full dealing with the repercussions of this crisis, which will likely slow efforts on other progressive priorities, and almost certainly make many of the concerns around biotechnology take a back seat to more pressing problems.

Still, one major goal of progressive bioethicists—universal health care—seems within reach. Obama's campaign focused heavily on his health-care plan. Even with the current economic problems, President Obama has said he will act swiftly to keep his campaign pledge to provide coverage to all Americans. Additionally, health-care advocates have vowed to continue putting pressure on Washington until needed changes are made to the system, and legislators seem genuinely encouraged about the prospects for serious reform.

President Obama has also taken quick action to repair our failed stem cell policy. While increasing science funding may prove difficult in the current financial climate, it is absolutely critical that research on embryonic stem cells receive federal support, particularly as state budgets that were funding a sizeable portion of the research come under increasing pressure to cut costs. And progressives will be further cheered if President Obama continues his commitment to scientific research and the use of data, letting empirical fact and not ideology drive regulatory and governmental decisions.

While it is almost certain that the Obama administration will be much more hospitable to progressive viewpoints than the Bush administration was, this does not mean that progressives have any easy task. In bioethics, we first have the serious issue of settling questions of values and policy within the progressive movement that have been raised by authors in this book, as well as others that will certainly arise. Progressives will have to undertake a serious discussion about the appropriate limits to be placed on new technologies, what resource allocation properly reflects the principles of social justice, and the ways in which bioethical expertise will be utilized in future policy decision-making processes. And simply having political power will not change the fact that concerns about many new biotechnologies cut across traditional political lines, creating tensions within the existing progressive movement and potentially leading to unexpected cooperation between normally opposing political camps. Additionally, as the commercial applications of biotechnology become more of a reality, progressives will need to reevaluate and confront their relationship with pharmaceutical companies and other private entities seeking to participate in and profit from the revolution in biology.

Despite these challenges, the future of progressive bioethics looks bright. While we face difficult times, we also have the chance to make significant and much needed changes. People realize that the same old ways of doing things are no longer effective, and are open to new possibilities and real reform. By adhering to core progressive values of social justice, critical optimism, and practical problem solving, progressives can help improve the world, ushering in a better future and harnessing change to benefit the common good.

About the Contributors

Sam Berger   is a J.D. candidate at Yale Law School. Before attending Yale, he worked as a researcher at the Center for American Progress, focusing on bioethics, civil rights and civil liberties, and the history of progressive thought.

Daniel Callahan   is a senior researcher and President Emeritus at the Hastings Center and a co-director of the Yale-Hastings Program in Ethics and Public Policy. He is the author of *Taming the Beloved Beast: How Medical Technology Costs Are Destroying Our Health Care System* (forthcoming) and a co-author of *Medicine and the Market: Equity v. Choice* (2006).

Arthur Caplan   is the Emmanuel and Robert Hart Professor of Bioethics and director of the Center for Bioethics at the University of Pennsylvania in Philadelphia. He is the author or editor of 29 books and over 500 papers in refereed journals of medicine, science, philosophy, bioethics, and health policy. His most recent books are *Smart Mice, Not So Smart People: An Interesting and Amusing Guide to Bioethics* (2006) and (as co-editor) *The Penn Center Guide to Bioethics* (forthcoming).

R. Alta Charo   is the Warren P. Knowles Professor of Law and Bioethics at the University of Wisconsin at Madison, where she is on the faculties of the Law School and the Medical School's Department of Medical History and Bioethics. Author of nearly 100 articles, book chapters, and government reports, she serves on the editorial boards of the *Journal of Law, Medicine and Ethics, Cloning: Science and Policy, Public Library of Science—Medicine*, and the *Monash Bioethics Review*. From 1996 to 2001 she was a member of President Clinton's National Bioethics Advisory Commission.

Marcy Darnovsky   is the associate executive director of the Center for Genetics and Society, an Oakland-based public affairs organization working to encourage responsible uses and effective societal governance of new reproductive and genetic technologies.

John H. Evans   is an associate professor of sociology at the University of California, San Diego. He is the author of *Playing God? Human Genetic Engineering and the Rationalization of Public Bioethical Debate* (2002) and a co-editor of *The Quiet Hand of God: Faith-Based Activism and the Public Role of Mainline Protestantism* (2002).

Kathryn Hinsch is the founder and board president of the Women's Bioethics Project, a non-profit, non-partisan public policy institute based in Seattle.

James Hughes is a bioethicist and a sociologist at Trinity College in Hartford. He serves as the executive director of the Institute for Ethics and Emerging Technologies and produces the IEET's syndicated weekly radio program Change-surfer Radio. He is the author of *Citizen Cyborg: Why Democratic Societies Must Respond to the Redesigned Human of the Future* (2004).

Richard Lempert is the Eric Stein Distinguished University Professor of Law and Sociology Emeritus at the University of Michigan. A co-author of *A Modern Approach to Evidence* (2000) and *An Invitation to Law and Social Science* and a co-editor of *Under the Influence: Drugs and the American Workforce* (1994), he is currently the Basic Research Deputy and Chief Scientist in the Human Factors/Behavioral Sciences Division of the Science and Technology Directorate of the Department of Homeland Security.

William F. May is a fellow at the Institute of Political Ethics and Public Life at the University of Virginia. In 2007 he held the Maguire Chair in American History and Ethics at the Library of Congress. Author of *The Physician's Covenant* (revised edition, 2000) and *Beleaguered Rulers: The Public Obligation of the Professional* (2001), he served on the Clinton Task Force on Health Care Reform in 1993 and on the President's Council on Bioethics in 2002–2004.

Eric M. Meslin is the director of the Indiana University Center for Bioethics, the Associate Dean for Bioethics, a professor of medicine, and a professor of medical and molecular genetics at the Indiana University School of Medicine. He is also a professor of philosophy in the School of Liberal Arts and a co-director of the IUPUI Signature Center Consortium on Health Policy, Law, and Bioethics. He served as executive director of the National Bioethics Advisory Commision under President Clinton. He has more than 100 publications, including *Belmont Revisited: Ethical Principles for Research with Human Subjects* (2005, co-edited with James Childress and Harold Shapiro).

Jonathan D. Moreno is the David and Lyn Silfen University Professor of Ethics and a professor of medical ethics and of history and sociology of science at the University of Pennsylvania; he also holds a secondary appointment as a professor of philosophy. He is a senior fellow at the Center for American Progress in Washington, where he edits the magazine *Science Progress*. He co-chaired the Committee on Guidelines for Human Embryonic Stem Cell Research and has served as a senior staff member for two presidential advisory commissions. His books include *Mind Wars: Brain Research and National Defense* (2006), *In the Wake of Terror: Medicine and Morality in a Time of Crisis* (2003), and *Undue Risk: Secret State Experiments on Humans* (2001).

Michael Rugnetta is the fellows assistant to Jonathan Moreno at the Progressive Bioethics Initiative. He graduated from the University of Pennsylvania in May 2007 with a double major in political science and cognitive science.

Harold T. Shapiro is president emeritus and a professor of economics and public affairs in the Department of Economics and the Woodrow Wilson School of Public and International Affairs at Princeton University. In 1996 he was appointed by President Clinton to chair the National Bioethics Advisory Commission. In 1990 he was appointed to serve as a member and vice chair of President Bush's Council of Advisors on Science and Technology. Co-editor of *Universities and Their Leadership* (1998), he currently chair of the board of DeVry Inc.

Paul Root Wolpe is the Asa Griggs Candler Professor of Bioethics and the director of the Center for Ethics at Emory University. He is also a co-editor of the *American Journal of Bioethics*, a past president of the American Society of Bioethics and Humanities, and the bioethics advisor to the National Aeronautics and Space Administration and the Planned Parenthood Federation of America.

Laurie Zoloth is a professor of medical humanities, bioethics, Jewish studies, and religion at Northwestern University. She is also the director of the Northwestern University Center for Bioethics, Science and Society and of the Brady Program for Ethics and Civic Life in Northwestern University's Weinberg College of Arts and Sciences. A former president of the American Society for Bioethics and Humanities, she received the Society's award for Distinguished Service to the Field in 2007. She is the author or editor of a number of books, including *The Human Embryonic Stem Cell Debate: Science, Ethics, and Public Policy* (2001).

Index

Basic Bioethics

Glenn McGee and Arthur Caplan, editors